*Ions, Electrodes,
and Membranes*

Ions, Electrodes, and Membranes

Jiří Koryta

J. Heyrovský Institute of Physical Chemistry and Electrochemistry,
Czechoslovak Academy of Sciences,
Prague

A Wiley–Interscience Publication

JOHN WILEY & SONS
Chichester · New York · Brisbane · Toronto · Singapore

Library of Congress Cataloging in Publication Data:

Koryta, Jiří.
  Ions, electrodes, and membranes.

  'A Wiley–Interscience publication.'
  Includes bibliographical references and index.
  1. Electrochemistry.   I. Title.
QD553.K673   541.3'7   81-14762
ISBN 0 471 10007 2 (Cloth)   AACR2   12|8|82
ISBN 0 471 10008 0 (Paper)

British Library Cataloguing in Publication Data

Koryta, Jiří
  Ions, electrodes, and membranes.
  1. Electrochemistry
  I. Title
541.3'7   QD553

ISBN 0 471 10007 2 (Cloth)
ISBN 0 471 10008 0 (Paper)

Filmset by Speedlith Photo Litho Ltd., Manchester and printed by Vail-Ballou Press Inc., Binghamton, N.Y., U.S.A.

Contents

Preface

The present book is an introduction to electrochemistry destined particularly for people coming from other fields of science. Electrochemistry is one of the oldest branches of physical chemistry and, in fact, of chemistry at all. The first great electrochemist was the Italian Luigi Galvani (1737–1798) who made his main discovery as early as 1791! In spite of this it took a considerable time, about three-quarters of a century, before two large regions formed in electrochemistry: one of them dealing with ion-containing liquids and the other one concerned with processes at electrodes. Later, particularly under the influence of the study of living cells, a new region emerged—the study of membranes. As I expect that at present this field has greatest chance for important discoveries, I reserve a special chapter for it.

In view of its great age, electrochemistry has an entrenched kind of language which makes its study rather difficult, both for a biologist and a physicist. Therefore I have tried to approach each problem in a simple physical manner keeping in mind that electrochemistry is an experimental discipline. At the beginning I always describe a simple (sometimes rather idealized) experiment and only then formulate a more general law and discuss its consequences. In my opinion this is a more suitable way of explanation than to postulate a certain law of nature and only use the experiments to illustrate its consequences. Needless to say, the choice of detailed information shows how much I stress the importance of electrochemistry, particularly for biology.

The reader will find additional information on various topics in this book in the books and reviews quoted in the text, also in the textbooks on electrochemistry: D. A. McInnes, *The Principles of Electrochemistry*, Reinhold, New York, 1961; G. Kortüm, *Treatise on Electrochemistry*, Elsevier, Amsterdam, 1965; J. O'M. Bockris and A. K. N. Reddy, *Modern Electrochemistry*, Plenum, New York, 1970; and J. Koryta, J. Dvořák, and V. Boháčková, *Electrochemistry*, Science Paperbacks, Chapman and Hall, London, 1973, and in three series of 'Advances', i.e. *Advances in Electrochemistry and Electrochemical Engineering* (ed. by P. Delahay, H. Gerischer, and C. W. Tobias), Wiley–Interscience, New York; *Modern Aspects of Electrochemistry* (J. O'M. Bockris and B. E. Conway, eds.), Butterworths, London, and Plenum, New York; and *Electroanalytical Chemistry* (A. J. Bard, ed.), M. Dekker, New York. Furthermore, there are three compendia of electrochemistry, C. A. Hampel (Ed.), *The Encyclopedia of Electrochemistry*, Reinhold Publishing Co., New York, 1964, A. J. Bard (Ed.), *Electrochemistry of Elements* (a series in publication since 1973), Marcel Dekker, New York, and J. O.'M. Bockris, B. E. Conway and E. Yeager (Eds.), *Comprehensive Treatise of*

Electrochemistry (a ten volume series in publication since 1980), Plenum Press, New York.

I am much indebted to Dr. A. Ryvolová-Kejharová, Mrs. D. Tůmová, and my wife for their help with the manuscript and to Mrs. Kozlová for drawing the figures. Dr. M. Hyman-Štulíková kindly revised the English.

Chapter 1

Ions

Origin

A simple experiment will demonstrate how ions are formed: gaseous sodium can be obtained by evaporation of metallic sodium in an evacuated tube at a sufficiently high temperature (above 883°C). The radiation from a light bulb is then passed through this gas and a spectroscope is used to observe the line absorption spectrum of sodium (Fig. 1). Each line corresponds to the radiation energy necessary for transferring the outer electron of sodium to a higher energy level after absorbing a photon with an energy $h\nu$. The Planck constant, which is a universal physical constant, is denoted by h and the frequency of light waves corresponding to the line in the absorption spectrum by ν. Close examination of the spectrum reveals a conspicuous feature, i.e. the distances between the lines gradually become smaller in the direction toward higher frequencies (or smaller wave-lengths) and finally reach a limiting value, called the edge of the spectrum. The energy corresponding to the edge of the spectrum is just sufficient for tearing-off the outer electron to form the sodium ion

$$Na \rightarrow Na^+ + e$$

This process is called ionization and can be brought about by a number of other mechanisms. For example, a high-energy photon can knock out one of the inner electrons of the sodium atom which then restores the stable structure with

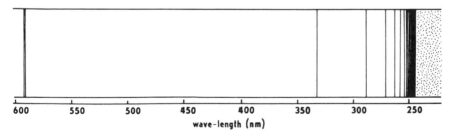

Fig. 1. The absorption spectrum of gaseous sodium. The adsorption lines are ordered according to their wave-length λ in nanometres. The frequency is $\nu = c/\lambda$, where c is the velocity of light in a vaccum. The line doublet on the left-hand side corresponds to the intense yellow colour in the emission spectrum of sodium. On the right-hand side the lines become denser until the edge of the series of lines is reached, corresponding to the ionization of sodium

occupied K and L shells by regrouping the other electrons. The final structure is again the Na$^+$ ion. Ionization can also be caused by collision of the atom with an elementary particle (electron, neutron, proton, etc.) or with other ions or atoms, etc. Ions are formed from molecules in the same way as from atoms. Negatively charged ions originate from reactions with an electron (for example, $O_2 + e \rightarrow O_2^-$).

There are simpler, more "chemical" methods of forming ions. Consider feeding gaseous hydrogen chloride, which consists of HCl molecules, into water. The resulting electrically conducting solution of hydrochloric acid contains exclusively chloride ions Cl$^-$ and hydrogen ions (which we shall provisionally denote as H$^+$) and virtually no HCl molecules. In contrast to pure water, this solution is an excellent conductor of electricity. During dissolution of HCl in water the reaction

$$HCl \rightarrow H^+ + Cl^-$$

obviously took place. Why could this happen so easily—at room temperature and without the influence of radiation? The answer can be found in the effect of the solvent. The solvent molecules form a stabilizing *hydration sheath* around each ion. The ionization of hydrogen chloride is followed and, in fact, made possible, by hydration of the ions formed from HCl. In general, for an arbitrary solvent, this process is called *solvation*. The reaction of ionization of HCl is accompanied by a further process

$$H^+ + Cl^- + xH_2O \rightarrow H^+.aq + Cl^-.aq$$

where the symbol aq indicates that the ions are solvated.

A similar electrically conductive solution consisting of hydrated ions is obtained by dissolving solid sodium chloride in water. This is somewhat different from dissolving hydrogen chloride in water. It has been proven by various physical measurements that the crystal lattice of sodium chloride consists of the sodium and chloride ions. The regular structure of rock salt, a typical ionic crystal (see Fig. 2), is a result of electrical forces attracting the Na$^+$ cations and Cl$^-$ anions and repulsing ions with the same sign (Cl$^-$ and Cl$^-$, for example) and the electron sheaths of all the ions. This interaction determines the effective dimensions of the ions. The effective crystallographic ionic radii given by Goldschmidt and somewhat modified values from Pauling are listed in Table 1. During dissolution the ions are only set free from the lattice and hydrated (see Fig. 3).

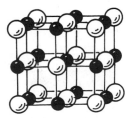

Fig. 2. The structure of the cubic crystal lattice of sodium chloride. The sodium ions are represented by black spheres while the chloride ions correspond to white spheres

Table 1 Comparison of Pauling's ionic radii with Goldschmidt's values G in nanometers

		Li	Be^{2+}		
		0.06	0.031		
		G0.078	0.034		
O^{2-}	F^-	Na^+	Mg^{2+}	Al^{3+}	Si^{4+}
0.14	0.136	0.095	0.065	0.05	0.041
G0.132	0.133	0.098	0.078	0.057	0.039
S^{2-}	Cl^-	K^+	Ca^{2+}	Sc^{3+}	Ti^{4+}
0.184	0.181	0.133	0.099	0.081	0.068
G0.174	0.181	0.133	0.106	0.083	0.064
Se^{2-}	Br^-	Rb^+	Sr^{2+}	Y^{3+}	Zr^{4+}
0.198	0.195	0.148	0.113	0.093	0.08
G0.191	0.196	0.149	0.127	0.106	0.087
Te^{2-}	I^-	Cs^+	Ba^{2+}	La^{3+}	Ce^{4+}
0.221	0.216	0.169	0.135	0.115	0.101
G0.211	0.22	0.165	0.143	0.122	0.102

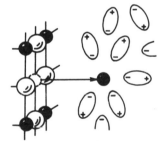

Fig. 3. The sodium ion leaves the crystal lattice of NaCl and is solvated by water molecules

Conduction of Electric Current

If two metal plates are placed parallel in a vacuum and connected to a d.c. source, then the electric current forms a positive charge on one of the plates and a negative charge on the other one, but no charge is transported between the plates—only the voltage (or, as it will be called here, the electrical potential difference) increases. Such a device is called *a condenser*, since it helps the charge to become "denser" on the plates.

The ratio of the charge brought to the positive plate of the condenser and the resulting electrical potential difference ΔV is called the capacity of the condenser C

$$C = Q/\Delta V \qquad (1.1)$$

When the plates of the condenser are connected to a low-voltage source, like an automobile storage battery (voltage 12 V) or to an electric grid (the a.c. voltage has to be first transformed to d.c. by a rectifier), the condenser is charged instantaneously and the current flowing to the plates rapidly decreases. When the condenser is connected to a high-voltage source like a Wimshurst machine or van de Graaf generator, an electric discharge appears between the plates. The electrons jump from the negative to the positive plate and an electric current starts to flow.

In a subsequent experiment the space between the plates is filled with a substance having a high *electric dipole*[1] like, for example, 1-chloropropane $CH_3CH_2CH_2Cl$. Chlorine is an element with high electronegativity (the tendency to attract electrons), therefore the end of the molecule carrying a chlorine atom exhibits a slightly excessive negative charge, while the other end becomes slightly positive. As a whole the chloropropane molecule is electroneutral.

The electric field which exists in the space between the plates of a charged condenser acts on the chloropropane molecules so that their positive ends are deflected somewhat towards the negative plate and the negative ends in the opposite direction. Complete orientation along the lines of force of the electric field is prevented by the thermal motion of the molecules—in a week electric field the dipoles are aligned only slightly. The electric field also influences the distribution of charge inside some molecules to form new dipoles. The dipoles originally present in the molecule are called *permanent* while the new ones are *induced*. The dipoles partially compensate the electric effect of the charge present on the plates and the electric field (and thus also the voltage) decreases. Equation (1.1) indicates that, at constant charge Q and decreased voltage ΔV, the capacity C increases.

When the space between the plates is filled, for example with paraffin (which is a mixture of long-chain hydrocarbons), the capacity of the condenser increases very little. Such a substance is *non-polar* and the space inside the condenser behaves almost in the same way as in a vacuum.

Any type of matter which prevents an electric current from flowing between the plates of a condenser is called a *dielectric* (including a vacuum); the substances which have so far replaced the vacuum in the condenser are termed *insulators*. Various dielectrics increase the capacity of the condenser to a different degree. Thus, it seems suitable to express this ability quantitatively. For this purpose a quantity called the *relative dielectric constant* or, better, the *relative permittivity*, denoted D, is used. The relative permittivity indicates the degree to which a given dielectric increases the capacity of the condenser compared with a vacuum condenser with the same size and distance between the plates. The values of the relative permittivities of some liquids are listed in Table 2. The relative permittivity decreases with increasing temperature and is almost independent of the electric field (for weak fields).

The relative permittivity appears in the equation for the capacity of the plate condenser

$$C = (D\varepsilon_0 S)/d \qquad (1.2)$$

where ε_0 is the permittivity of a vacuum, $\varepsilon_0 = 8.85 \times 10^{-12}\,\mathrm{Fm^{-1}}$, S is the area of one plate, and d is the distance between the plates.

Table 2   Relative permittivities of selected solvents

Solvent	D	°C	Solvent	D	°C
Water	78.4	25	Ethanol	24.6	25
(Ice)	94	−2	1-Butanol	17.5	25
Formamide	109	20	Terc-Butanol	2.7	25
HCN (liq.)	95	21	1,2-dichloro-		
			benzene	9.8	20
Formic acid	47.9	18.5	Chloroform	4.8	20
Dimethyl-			Tetrachloro-		
sulphoxide			methane	2.23	20
	46.7	25	Benzene	2.23	20
Acetonitrile	37.5	20	Dioxane	2.28	20
Dimethylform-					
amide	36.7	25	Pentane	1.8	20.3
Nitrobenzene	36.1	20	Ammonia (liq.)	21	−34
Methanol	32.7	25			

The effect of replacing the dielectric by a piece of metal touching both plates is not surprising. After the plates come into contact, an electric current starts to flow through the metal—the charge cannot be retained on the plates because they are conductively connected. Thus, metal is an *electric conductor*.

There are various kinds of conductors. They may be distinguished according to the type of electric particles which carry the current through them. In metals these particles are electrons—metals are electronic conductors. In the crystal lattice of a metal the metal atoms are ionized, the electrons are set free and the electron clouds (orbitals) overlap in such a way that all the electrons are common to all the ions. Under these conditions the electrons in the metal behave in a way similar to molecules in a gas. (This follows from Fermi–Dirac statistics governing the behaviour of electrons in solids.) Thus, we speak of an 'electron gas' where the probable position of an electron changes in the same irregular manner as the position of a molecule in a gas, where it is constantly under the influence of collisions with other molecules.

Like electrons in an isolated atom, the electrons in a metal follow Pauli's principle which states that an energy level in the system can be filled by a maximum of two electrons with different spins.[2] In a metal, the energy levels of the electrons differ very little from each other. Thus, they form a continuous band which we call the *conductivity band* (Fig. 4). The probability of an electron

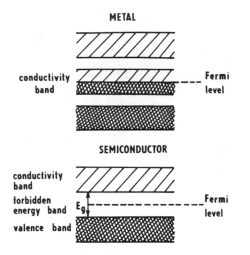

Fig. 4. The band structure of a metal and of a semiconductor. Simply hatched areas are not occupied but can be occupied by electrons while void areas are not accessible to electrons. Double hatched areas are occupied by electrons. The vertical coordinate of the scheme corresponds to the energy of the electrons. E_g denotes the energy difference between the upper and the lower edge of the forbidden energy gap (the gap width). In a metal half of the conductivity band is filled with electrons. In a semiconductor the conductivity band is empty. Between this and the valence band the forbidden energy gap is situated; the Fermi level lies in the middle of the gap

occupying a level with energy ε is given by the Fermi function

$$P = \frac{1}{1 + \exp\left[(\varepsilon - \varepsilon_F)kT\right]}$$

where ε_F is the energy of the Fermi level. This level has the probability of occupation exactly one-half. k is the Boltzmann constant and T is the absolute temperature (for details, see Fig. 5).

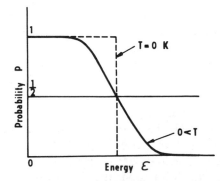

Fig. 5. The Fermi function for the energy distribution in a metal. ε_F denotes the energy of the Fermi level. At a temperature of absolute zero (0 K), the energy levels below ε_F are completely filled while those above ε_F are empty. At higher temperatures the levels in the neighbourhood of ε_F are only partly filled

We already know that the electrons in the metal move in various directions and with various velocities. Connection of a voltage source to a metallic conductor will result in a slight change in the motion of the electrons—the component of the velocity in the direction of the field (i.e. from the negative to the positive pole) will prevail somewhat over the component in the opposite direction. This preference of the velocity in a certain direction is very small compared with the absolute magnitudes of the velocities of the irregular motion, but is just sufficient to cause the flow of an electric current. This is a typical property of many physical and chemical processes. They proceed in various directions with high velocities but in the equilibrium state all these 'partial' processes cancel each other out and no change is observed. When a partial process in one direction is slightly faster than in the opposite direction, a change in the physical or chemical system becomes apparent (current flow in the present case). Under these circumstances the rate of change (here the flow of electric charge, i.e. electric current) is proportional to the generalized force which causes the change (applied voltage or electric potential difference in the present case). Thus, we have deduced Ohm's Law in a simple intuitive way, stating that electric current I is proportional to electric potential difference V

$$I \sim \Delta V$$

The proportionality constant is the conductance G which is equal to the reciprocal of the resistance R.

As mentioned above, the charge carriers in metallic conductors are electrons, while metal ions in the lattice do not transfer electric charge because they occupy fixed positions. These positions are not in fact completely fixed because they vibrate around average lattice positions. The frequency as well as the amplitude of these vibrations increases when the temperature of the metal increases. The more the lattice ions vibrate, the more they hinder the free movement of the electrons, so that the resistance of a metallic conductor increases with increasing temperature.

When a piece of very pure elemental germanium is placed between two metal plates (the purification of such materials has been highly perfected during the past few decades) and a voltage source is connected to the plates, a very tiny electric current starts to flow. This current increases when the temperature is increased. In appearance, germanium is much like a metal, being grey with a metallic sheen, but it conducts electricity very imperfectly. It belongs among the *semiconductors*.

The electronic structure of a semiconductor is different from that of a metal. The energies of the outer electrons in the electron sheath of the atoms in the semiconductor almost completely fill the valence band (Fig. 4). Above this band lies the conductivity band, which is almost empty. There is no energy level which could be occupied by the electrons of the semiconductor in the energy interval between these two bands. This property is described by the appropriate term 'forbidden energy gap'. The thermal vibration of the electrons can be so great that they can 'jump up' to the conductivity band and participate in the conduction of electricity. When the temperature is increased, more electrons reach the conductivity band, so that the conductivity of a semiconductor increases with increasing temperature. The electron which has 'jumped' up from the valence band has added a negative charge to the conductivity band. According to the law of conservation of charge, when a charge of a definite sign is formed, another charge of opposite sign must emerge at the same time. Thus, the electron has left a vacant position in the valence band—a hole or vacancy—with a positive charge. 'Chemically' speaking this means that an electron has been set free from a germanium atom and can freely move in the crystal lattice, while the resulting Ge^+ ion can exchange its positive charge for an electron from the neighbouring atoms.

If an electric field is applied to pure germanium, the electrons migrate in one direction and the holes in the opposite direction (again the velocity of the electrons or holes in this particular direction slightly prevails over the velocities in the other directions). Such 'intrinsic' semiconductors conduct electric current poorly and are technologically relatively unimportant. A revolution in electronics, marked by the progress from simple rectifying diodes to microprocessors with integrated circuits, was based on the discovery of semiconductors with artificially implanted impurities, i.e. of semiconductors doped with minute quantities of admixtures.

A very small amount of arsenic can be added to pure germanium, leading to the following 'chemical' process: an electron is split off from the outer shell of the electron sheath of arsenic and is added to the neighbouring germanium atoms in such a way that it can move freely in the germanium lattice. Because of its ability to donate electrons, arsenic is termed an 'electron-donor'. In the band scheme of germanium this phenomenon is demonstrated as shown in Fig. 6. The admixture of arsenic brings about a new donor energy level just below the conductivity band. The electrons need only a small amount of energy E_d to move up to the conductivity band, thus increasing the conductivity of germanium. The semiconductor that contains a donor admixture and has conductivity based on the motion of electrons (negative charge carriers) is termed an n-type semiconductor.

The opposite process occurs when germanium is doped with boron instead of with arsenic. Boron is an electronegative element with a tendency to complete its electron shell. It receives one electron from germanium and is converted to the negative ion B^-. At the same time, a hole is formed in the electron structure of germanium. This situation is depicted in Fig. 6. Boron is an 'electron-acceptor' and, when present as an admixture in germanium, gives rise to a new acceptor energy level just above the valence band. The transfer of an electron from the valence band to the acceptor band requires a small energy E_d. The hole remaining in the valence band contributes to the conductivity of the semiconductor. Because it is a positive charge carrier, this material is called a p-type semiconductor. Increasing the temperature has the same influence on doped semiconductors as on intrinsic semiconductors, because the thermal energy of the electrons permits them to reach the donor or acceptor levels.

In the next experiment, two plates made of a certain metal, for example, copper,

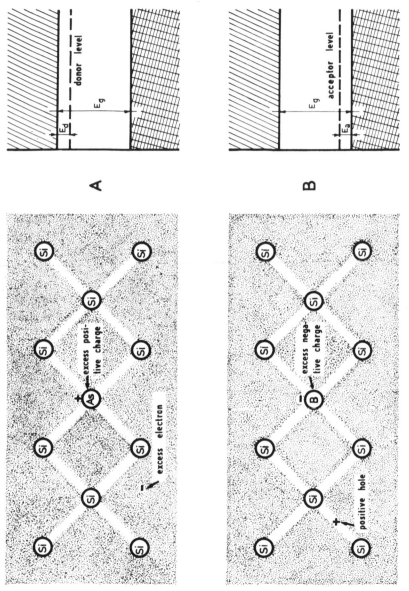

Fig. 6. Schematic representation of a semiconductor (germanium) with impurities. Doping with an electropositive element (As) results in a new 'donor' level immediately below the conductivity band. The electrons jumping from this level to the conductivity level induce n-type conductivity in the semiconductor (the charge carriers are the electrons with a negative charge). By doping with an electronegative element (B) a new 'acceptor' level is formed just above the valence band. The electrons jumping from the valence band into this new band leave holes (positive charge carriers) in the valence band. Thus, the material becomes a p-type semiconductor. From C. Kittel

are immersed in a copper sulphate solution. When a source of d.c. voltage is connected to the plates, an electric current starts to flow, accompanied by chemical changes at the copper plates, which will be termed *electrodes*.

The copper plate through which the positive charge from an external source of electricity enters the solution is dissolved. This process can be described by the equation

$$Cu(metal) \rightarrow Cu^{2+}(solution) + 2e(metal) \qquad (1.3)$$

In fact, copper ions were already present in the metal lattice, so that equation (1.3) can be broken down into several partial steps

$$Cu(metal) \rightleftarrows Cu^+(metal) + e(metal)$$
$$Cu^+(metal) \rightarrow Cu^+(solution)$$
$$Cu^+(solution) \rightarrow Cu^{2+}(solution + e(metal)$$

Metallic copper is deposited on the other electrode, where the electrons enter the solution, apparently in the process

$$Cu^{2+}(solution) + 2e(metal) \rightarrow Cu(metal) \qquad (1.4)$$

This reaction again takes place as a sequence of partial steps

$$Cu^{2+}(solution) + e(metal) \rightarrow Cu^+(solution)$$
$$Cu^+(solution) \rightarrow Cu^+(metal)$$
$$Cu^+(metal) + e(metal) \rightleftarrows Cu(metal)$$

The last step must always involve transfer of electrons to the metal because the electrode must remain electroneutral (even if copper remains in the ionic form in the lattice).

The electrode which accepts a positive charge from the solution or injects a negative charge into the solution is termed the *cathode*, while the other electrode injecting a positive charge into the solution or accepting a negative charge from the solution is the *anode*.

Close examination of equations (1.3) and (1.4) reveals that the substances present in the solution which are formed at the electrodes as a result of the flow of electric current through the solution, or from which the final products originate, are exclusively ions. We can assume (and other experiments have justified our doing so) that only ions participate in the conduction of electric current in the $CuSO_4$ or, in other words, only ions function as *charge carriers*. The conductor where electricity is communicated by the ions is termed an *electrolyte*.

The whole process described above is called *electrolysis*. This term, as well as *electrode*, *anode*, *cathode*, and *electrolyte*, were introduced by M. Faraday in 1835 (see Appendix A).

Electrolysis can be separated into two principal processes:

(1) The processes taking place directly at the electrodes which are connected with charge transfer between the electrode and the electrolyte solution.

(2) Charge transfer in the bulk of the solution called ion migration (Latin *migro* = wander). In addition, matter is also transported by other processes in the solution (see p. 45). The processes taking place directly at the electrodes are dealt with in Chapter 2.

The ions that are present in the cupric sulphate solution are Cu^{2+} cations and SO_4^{2-} anions. At higher concentrations $Cu(SO_4)_2^{2-}$ ions become noticeable, but will be neglected here as the concentration used in our experiment will not be high.

Imagine that the charge in the electrolyte is transferred only by the motion of the Cu^{2+} cations and that the SO_4^{2-} anions remain in fixed positions. Then, during current flow, the $CuSO_4$ concentration would remain constant at any point in the solution. However, in an actual experiment, the concentration around the electrodes increases at the anode and decreases at the cathode as a result of the mobility of both ionic species in the solution (in solution the Cu^{2+} cation moves more slowly than the SO_4^{2-} anion, but this is not important here). Therefore, when the Cu^{2+} ions are discharged at the electrode, new copper ions migrate to the electrode and, at the same time, sulphate ions migrate from the electrode. However, the electroneutrality condition must be preserved in the solution so that the total charge of the cations and the anions must be equal in any, even very small, volume of the solution. If this condition were not observed in a region of the solution, a space charge would appear and the repulsing force between the ions would expel excess ions from that region. The electroneutrality condition is expressed mathematically by the equation

$$z_+ n_+ + z_- n_- = 0$$

where z_+ and z_- are the *charge numbers* of the cation and of the anion, respectively, and n_+ and n_- are the numbers of cations and anions, respectively, in a volume of the solution (the exact definition of concentration can be found on p. 20). The charge number is defined as the ratio of the charge of the ion to the proton charge (for example, for $CuSO_4$, $z_+ = 2$, $z_- = -2$). Thus, when the sulphate ions migrate from the neighbourhood of the cathode, the total concentration of $CuSO_4$ there must decrease in order to preserve electroneutrality. On the other hand, the concentration of copper ions at the anode increases as a result of anodic dissolution of copper. Those ions migrate from the electrode and new sulphate ions approach it. In this way the electroneutrality is retained, leading to an increase in the copper sulphate concentration in the neighbourhood of this electrode.

The same properties as those exhibited by a copper sulphate solution are characteristic for solutions of innumerable salts, acids, and bases (see p. 30). However, the processes taking place directly at the electrodes are often quite complicated and differ, for example, when the electrolysis is carried out with electrodes made of different materials. The situation at the electrodes differs considerably from that in the bulk of the solution, where ion transport proceeds. Fortunately, it is possible to separate the processes at the electrodes from those taking place 'inside' the electrolyte solution and thus to elucidate both groups of phenomena. This separation was not apparent to the founders of electrochemistry such as Davy, Faraday, Grotthus and others[3] in the first half of the nineteenth century, and their views re-emerge with remarkable perseverance even now.

There are, of course, many more different types of electrolytes. For example, two metal plates connected to a d.c. voltage source can be fixed to a sodium chloride crystal. Although the crystal is composed of chloride and sodium ions, no current starts to flow. Lattice forces hold the ions in their positions and the electric field, if not extremely high, cannot remove them from these sites.

Thus, solid sodium chloride is a non-conductor (*insulator*). As a substance, though, it is also called an electrolyte because dissolution in water forms an *electrolyte solution*. Obviously, a substance which is an electrolyte need not immediately to be a conductor, but must achieve this property after a certain procedure. Here, the sodium chloride will not be dissolved but the temperature will be raised above 800°C. In this way we obtain an excellently conductive liquid, the *fused electrolyte* (or the ionic liquid).[4] Fused electrolytes are rather important; for example, the slag in metallurgical processes is a silicate-type fused electrolyte and aluminium is produced by dissolving aluminium oxide (obtained from the mineral bauxite) in fused cryolith Na_3AlF_6 and by electrolysis of this melt.

When, instead of sodium chloride, a crystal of silver chloride or bromide is placed between the metal plates,* we find that a distinct current starts to flow between the plates. This current is much smaller than that in a fused electrolyte or an aqueous electrolyte solution, but sufficiently distinguishes AgCl or AgBr from insulators like solid sodium chloride. Silver chloride or bromide are *solid electrolytes*.[5] In their crystal lattice the halide (chloride or bromide) ions are immobile while the silver ions are genuine charge carriers. Because of imperfections of the crystal some silver ions lie outside regular lattice positions (they are at *interstitial* positions, see Fig. 7) and some lattice positions are not

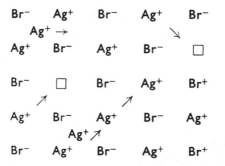

Fig. 7. Charge transfer in solid AgBr. The conductivity is due to interstitial ions, Ag^+, and ionic holes. According to J. I. Frenkel

---

* Silver plates must be used which are brazed to the crystal with a silver solder.

occupied (these free positions are termed ionic *vacancies* or *holes*). The transfer of silver ions to and from interstitial positions and vacancies enables the transport of charge in the crystal. The conductivity of solid electrolytes rises as the temperature increases. Solid electrolytes are important for ion-selective electrodes (see Chapter 3, p. 137) and for certain types of fuel cells (see Chapter 2, p 87).

\For chalcogenides (sulphides, selenides, tellurides) of silver, lead, cadmium, etc. *mixed conductivity* is typical. In these materials the electric current is carried by ions as well as by electrons. The relative contributions of the two types of charge carrier depend, in addition to other features, on the deviation of the composition of the crystal from the stoichiometric component ratio (for example, if the Ag_2S crystal contains a small concentration of excess silver or sulphur).\

When a gas composed of diatomic molecules is heated to a very high temperature, the molecules dissociate to atoms and at still higher temperatures the gas turns to ions, both positive and negative, and to electrons. The resulting system is a *plasma*, which cannot be termed a gaseous electrolyte because it is characterized by mixed conductivity—both ions and electrons act as charge carriers.

Finally, the strange solution formed by dissolving, e.g. metallic sodium in liquid ammonia, will be considered. The blue liquid deepens in colour when the concentration of sodium is increased and above 4 mol Na/l the liquid becomes grey with a metallic lustre. Initially, the conductivity increases roughly in direct proportion to the sodium concentration, and when the liquid acquires a metallic appearance it increases rapidly (see Fig. 8). The peculiar behaviour of these solutions is a result of the formation of solvated electrons in the liquid

$$Na(metal) + solvent \rightarrow e(solution) + Na^+(solution)$$

The electrons are located in cavities between the solvent molecules and are very reactive (for example, they act as strong reductants in many organic reactions). When their concentration increases, electron pairs, 'dimers', are formed. They posess anti-parallel spins (the spin of one electron lies in the opposite direction to that of the other one). At still higher concentrations the orbitals of the electrons start to overlap in the same way as in a metal lattice. The electrons in a

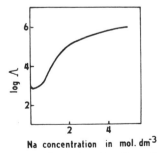

Fig. 8. Dependence of the molar conductivity of sodium dissolved in liquid ammonia on the Na concentration at −34°C. According to E. C. Evers

14

concentrated solution then form an electron gas, manifested by the change in appearance as well as by the striking increase in the conductivity.[6]

This phenomenon is often observed with alkali metal solutions in amine or other types of solvents. At present, the properties of such solutions in hexamethylphosphortriamide $[(CH_3)_2N]_3PO$ are being intensively studied.

References

1. P. Debye, *Polar Molecules*, Dover Publications, New York, 1945.
2. See, for example, C. Kittel, *Introduction to Solid State Physics*, John Wiley & Sons, New York, 1966.
3. See, for example, L. P. Williams, *Michael Faraday*, Chapman and Hall, London, 1965.
4. B. R. Sundheim (Ed.), *Fused Salts*, McGraw-Hill Book Company, New York, 1964. D. Inman and D. G. Lovering (Eds.), *Ionic Liquids*, Plenum Press, New York, 1981.
5. A. Huggins, 'Ionically conducting solid state membranes', in: *Advances in Electrochemistry and Electrochemical Engineering*, Vol. 10 (H. Gerischer and C. W. Tobias, eds.), Wiley–Interscience, New York, 1977. J. Hladik (Ed.), *Physics of Electrolytes*, Vols. 1 and 2, Academic Press, New York, 1972.
6. E. C. Evers, *J. Chem. Educ.*, **38** (1961) 590.

Ion Solvation

A small quantity of sodium chloride can be dissolved in water to form a very dilute solution. Compared to the original situation with crystals of sodium chloride and water (at the same temperature), the solution is somewhat cooler. When the same amount of NaCl is dissolved in methanol so that the resulting solution has the same concentration, the cooling effect is greater.

To elucidate these phenomena, we shall resolve the whole process into several steps characterized by the different amounts of energy accepted or released by the system. We are justified in doing so because this will be a thermodynamic procedure following the Hess Law. This law states that the resultant energy change in a given system is independent of the steps through which the system proceeds in passing from the original to the final state. The following imaginary process can be considered (see Fig. 9). First the sodium chloride will be

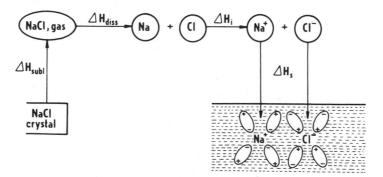

Fig. 9. The modified Born–Haber solvation scheme

evaporated into a vacuum. This step, connected with release of NaCl from the crystal lattice, requires the supply of the sublimation energy ΔH_{subl}.

All energy quantities appearing in this discussion are the *enthalpies*, i.e. the amount of heat supplied to or released from a system in a process taking place at a constant pressure. The symbol for the enthalpy is H (for details, see Appendix B). By convention, the energy accepted by the system is considered positive and the energy released negative.

The resulting gaseous NaCl will then dissociate into gaseous atomic sodium and chlorine. For this step the system must accept the necessary dissociation energy ΔH_{diss}. Then these gaseous atoms are transformed into ions

$$Na(gaseous) + Cl(gaseous) \rightarrow Na^+(gaseous) + Cl^-(gaseous)$$

Here the energy change ΔH_i, is given by the sum of the ionization potential of sodium I_{Na} and of the enthalpy of electron acceptance by chlorine $\Delta H_{e,Cl}$

$$H_i = I_{Na} + \Delta H_{e,Cl}$$

(The enthalpy of electron acceptance is sometimes called the electron affinity but, since in thermodynamics the term affinity is reserved for another type of energy quantity, it will not be used here.) Finally, these gaseous ions are simultaneously transferred into the solution where they interact, in various ways, with the solvent dipoles.

In fact, there exists an electrical potential difference between the bulk of the gaseous phase and of the solution. The ions then exert electrical work during their transfer. However, as cations and anions bearing identical charges of opposite sign are transferred, the electrical work for each of them is the same but with opposite sign. Therefore, the resulting total work during simultaneous anion and cation transfer across the electric potential difference between the two media is zero.

This process is generally called *ion solvation*[1] (for solvation in water the term *hydration* is used). Solvation is a process connected with the release of energy (an *exothermic* process) and therefore the solvation energy ΔH_s has a negative sign. The total energy change during dissolution of the NaCl crystal ΔH is given by the sum

$$\Delta H = \Delta H_{subl} + \Delta H_{diss} + I_{Na} + \Delta H_{e,Cl} + \Delta H_s$$

All the quantities on the right-hand side are positive (the corresponding processes are *endothermic*, i.e. they are connected with the acceptance of energy) except the solvation energy ΔH_s. The sum of the positive quantities is larger than the absolute value of the solvation energy $|\Delta H_s|$. Therefore the resulting energy change is positive so that, during the dissolution of NaCl, heat must be supplied from an external source or the temperature of the system drops (the internal energy of the system is consumed in the dissolution process).

In the dissolution in water and in methanol all the partial steps are the same, except solvation. The larger decrease in energy (the larger consumption of energy for the dissolution of NaCl) in methanol is a result of weaker solvation in this medium (i.e. lower absolute value of the solvation energy $|\Delta H_s|$).

The existence and extent of ion solvation is reflected in several other observations. For example, ions often move more slowly in solution than would be expected on the basis of their ionic radii determined by crystallographic measurements (cf. Table 1). The velocity of a spherical particle in a medium of viscosity η follows the Stokes Law

$$v = f/(6\pi\eta r)$$

where r is the radius of the particle. Assuming that the ion is a spherical particle, we can (very roughly) determine its radius from its mobility, which is evaluated on the basis of conductivity data as shown on p. 51. The difference between the radii determined from crystallographic and mobility data for small ions in water like lithium, which have a very large solvation sheath of water molecules, is striking. Another approach to the study of solvation is the measurement of the thermal capacity of the solution. This quantity depends, among other factors, on the ability of solvent molecules to move from one site to another (translational motion) and to vibrate and rotate. Under the influence of ions present in the solution, these abilities vary, so that the region of action of the ions on the solvent can be assessed by determining the capacity of the solution and by comparing it with the capacity of the pure solvent. On the basis of microwave spectra (for example, nuclear magnetic resonance, NMR) it has been found that, after dissolution of an electrolyte in water, some of the water molecules have a different frequency of vibration of the O—H bond than in pure water. These molecules obviously belong to the hydration sheath. It is not surprising that each of the three methods described, i.e. the study of ion mobility, of the heat capacity of the solution, and of the microwave spectra of the solvent, give different information on the extent of solvation, which is sometimes defined by the solvation number (the number of solvent molecules under direct effect of the individual ion), as in each method the influence of the ion on a different physical process is measured.

In order to comprehend the process of solvation the properties of the solvent alone will first be considered. In the liquid state molecules are a very small distance apart and are held together by forces like dipole–dipole interaction[2] and van der Waals attraction.[3] The van der Waals force is analogous to the interaction between the dipoles but acts between non-polar molecules. Because of vibration even in a non-polar symmetrical molecule, an asymmetrical charge distribution is statistically formed, thus inducing a polar distribution in neighbouring molecules, so that a kind of instantaneous dipole–dipole interaction results (Fig. 10). While the force of the dipole–dipole interaction is proportional to r^{-3} (r is the distance between the polar molecules) the van der Waals attraction is proportional to r^{-7}—a typical short-range interaction.

Pure van der Waals interactions are present in very simple liquids like liquefied inert gases (neon, argon, etc.) which are, unfortunately, of little interest for fields of chemistry and physics other than for the theory of the liquid state. In liquids that are often used as solvents, other forces simultaneously play an important

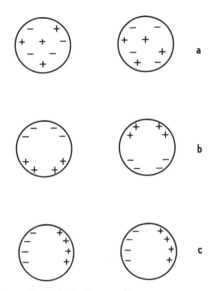

Fig. 10. The origin of van der Waals forces. On average, non-polar molecules have a charge distribution shown in (a) but at any arbitrary instant this distribution is assymmetrical so that an instantaneous dipole is formed (b,c). The van der Waals force depends on the mean square of the dipole moment (charge or the positive end of the dipole multiplied by the distance between the ends of the dipole) which is larger than zero for all molecules (or atoms)

role. Thus, the rather complicated properties of water[4] may be explained by interactions between dipoles of water molecules and by association of these molecules through hydrogen bonds (hydrogen bridges). As a result of this association clusters of water molecules are formed in liquid water. These clusters, which are reminiscent of the hexagonal structure of ice (see Fig. 11), contain different numbers of water molecules (in warmer water smaller clusters prevail). Between these clusters may be real disordered 'liquid' water. The clusters rapidly decompose, reform, and are formed again from disordered water but, if this change could be stopped for a moment, some sort of water crystals would be observed, as well as irregularly oriented molecules.

If an ion appears in the solvent, its charge and its structure influence the solvent structure in its surroundings. For example, when lithium chloride is dissolved in water the small lithium cation causes the water dipoles to orient with their negative (oxygen) end toward the centre of the ion.* In this way a layer of oriented water molecules is formed in the immediate neighbourhood of the cation, called the *primary hydration sphere* (see Fig. 12). Those water molecules align others in the surroundings and a cluster of water molecules is formed around the cation, like the instantaneous crystal structure in pure water. There are, however, some

---

* The O—H bond angle in a water molecule is 112°, the positive pole of the dipole is situated between the hydrogen atoms while the negative end lies on the oxygen atom.

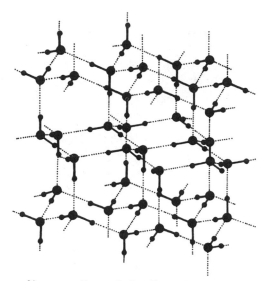

Fig. 11. The structure of ice according to L. Pauling. The large circles are oxygen atoms and the smaller ones are hydrogen atoms

differences because the molecules in the primary hydration sphere remain in place for a longer time than water molecules in a cluster that is not influenced by an ion charge. Thus, the cluster around the lithium cation is more stable than the usual cluster in water alone. Since the lithium ion reinforces the water structure, it is termed a *structure-making* ion. When in an aqueous solution the lithium ion is replaced by another alkali metal ion which has a large radius, such as the caesium ion, the result is different. Since this ion has a large electron sheath, the positive charge of the ion acting on the negative ends of the water dipoles is screened by the large electron envelope. At the same time, the electrons repulse the approaching negative ends of the dipoles so that the final result is a disordered

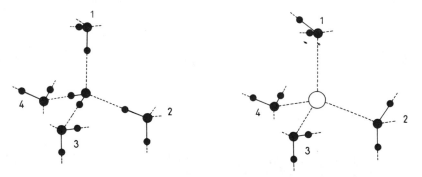

Fig. 12. The effect of ion solvation on the structure of water. In pure water the structures are similar to those in ice (the large full circles denote oxygen atoms, the small ones hydrogen atoms). By substitution of a water molecule by a cation (large full circle) four hydrogen bonds are disrupted and the oxygen atoms of the water molecules are directed towards the cation

distribution of water molecules similar to that in the disarrayed 'fluid' region of water. This property of caesium is termed '*structure-breaking*'.

Among inorganic ions, chlorides and fluorides are structure-making, and nitrates and perchlorates are structure-breaking. Ions which have no pronounced properties of one or the other kind are, for example, the potassium, bromide, and sulphate ions.

As shown above, solvation is obviously based on the interaction between the electric charge of the ion and the dipole of the solvent molecule. Substances with larger dipoles exhibit large solvation effects. Since more polar substances have a larger dielectric constant, they will solvate ions more strongly. This has already been made obvious from the comparison of solvation by water and methanol: water, with a dielectric constant about twice as large as methanol, exhibits stronger solvation.

Solvation is not observed only with ions. Polar groups in the molecule (in organic molecules, for example, $-C=O$, NH, $-NO_2$, etc.) can also be solvated as a result of dipole–dipole interaction of these groups with the solvent. The solvation of polar groups is important for chemical reactions where these groups participate and, again, decreases with decreasing polarity of the solvent.

When a substance with a typical non-polar group like a long-chain alkyl is dissolved in water, a peculiar phenomenon termed *hydrophobic interaction*[5] is often observed. Such a non-polar group is structure-making and produces an increase in the order of the system. In this way the entropy of the system would (under certain conditions, of course) decrease. Because natural processes tend to increasing disorder (i.e. to an increase in entropy—cf. Appendix B), the alkyl groups preferably join together forming complicated associates and leaving the remaining water in greater disorder. At the same time the water dipoles attract other water dipoles and repel less polar species from their environment. Because of these two effects, the non-polar groups are inclined to disappear from contact with the aqueous medium. This phenomenon is particularly significant for molecules containing a polar or ionized group in addition to the non-polar alkyl group (*amphiphilic* or amphipatic* molecules). They preferably enter the surface of a solution forming *monolayers* (Fig. 13A) or, at higher concentrations, spherical associates called *micelles* (Fig. 13B). If a molecule contains one polar group and two long-chain alkyl groups, such as various phospholipids, the hydrophobic interaction is so strong that a *bilayer film* is formed (see Chapter 3, p. 155).

References

1. O. Ya. Samoilov, *Structure of Aqueous Electrolyte Solutions and the Hydration of Ions*, Plenum Press, New York, 1965.
   B. Case, 'Solvation of ions', in: *Reactions of Molecules on Electrodes* (N. S. Hush, ed.), Wiley–Interscience, New York, 1971.

* Greek *amphi* = both.

20

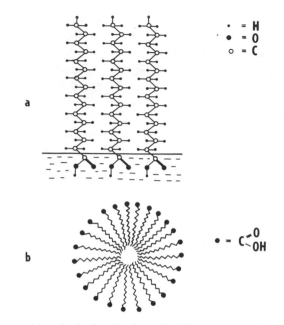

Fig. 13. Monolayers (a) and micelles (b) of amphiphilic molecules. At low concentrations, amphiphilic palmitic acid forms a monolayer at the water/air interface (a). At higher concentrations spherical structures, micelles, are formed (b)

2. R. W. Guerney, *Ionic Processes in Solution*, McGraw-Hill Book Company, New York, 1953.
3. K. Morokuma, *Accounts Chem. Res.*, **10** (1977) 294.
4. H. S. Frank, *Science*, **196** (1970) 635.
   F. Franks (Ed.), *Water*, Vols. I–V, Plenum Press, New York, 1972–1978.
5. C. Tanford, *The Hydrophobic Effect: Formation of Micelles and Biological Membranes*, 2nd Ed., John Wiley & Sons, New York, 1980.

Strong Electrolytes

In this section we consider solutions from another point of view. The properties of a solution depend strongly on the quantity of the substance dissolved. The representation of this quantity is expressed by various relationships between the amount of dissolved substance (the solute) and the amount of solution or solvent. Quantitative measures of these relationships are various *concentration scales*. Very often the representation of a solute is determined by the amount (in mols) of substance dissolved in a unit volume (usually dm^3 or litre) of the solution. In this way, the *molar concentration* is defined in units of $mol.dm^{-3}$. When a solution has the concentration m $mol.dm^{-3}$ we call it m-molar. The ratio of the amount of solute to the mass of the solvent is the *molality* (unit $mol.kg^{-1}$). The ratio of the amount of solute (in mols) to the total amount of all substances in the solution (in mols) is the *molar fraction* (a dimensionless quantity).

If a non-electrolyte (a substance which on dissolution does not dissociate into ions), e.g. cane sugar or saccharose, is dissolved in water and the solution is placed in an evacuated flask, then the gaseous phase above the solution contains only water vapour (at room temperature the evaporation of saccharose can be completely neglected). When the pressure of the water vapour is measured with a manometer it is found that, the more the concentration of saccharose increases, the lower is the vapour pressure. When the concentration of saccharose is less than about 0.1 mol.1^{-1} (the molar fraction is less than 0.00181), then the pressure of water vapour p obeys the equation

$$p_1 = x_1 p_0 \qquad (1.5)$$

where p_0 is the vapour pressure of pure water at the same temperature and x_1 is the molar fraction of water in the solution. This relationship can be advantageously described by the drop in the water vapour pressure in the presence of the dissolved substance (molar fraction x_2). As the molar fractions x_1 and x_2, according to the definition of molar fractions, are related by the equation

$$x_1 + x_2 = 1$$

equation (1.5) yields the relationship

$$(p_0 - p_1)/p_0 = x_2 \qquad (1.6)$$

Equivalent equations (1.5) and (1.6) are called the Raoult Law. Solutions that obey this law are termed ideal. They have the general property that the vapour pressure of any component of the solution is proportional to its content in the solution. In the form of equation (1.6) the Raoult Law is valid for non-electrolytes which differ in the concentration limits of validity of the law.[1]

The vapour pressure of a liquid depends on the number of molecules in the liquid which reach the surface of the liquid in a certain time interval and then have a chance of evaporation. Saccharose molecules reaching the surface cannot evaporate because the vapour pressure of saccharose is immeasurably low. Thus, saccharose molecules coming to the surface only decrease the probability of evaporation of water molecules, resulting in a decrease in the water vapour pressure.

A similar experiment is illustrated in Fig. 14. A vessel has two compartments, A and B. Compartment A contains only a pure solvent, while a solution of a non-electrolyte with molar fraction x_2 is in compartment B. Vertical tube T connected to compartment B is thin, so that its volume can be neglected with respect to the volume of the solution in B. Both A and B are connected by a porous semipermeable membrane which enables only the solvent molecules to move from one compartment to the other and prevents the solute molecules from passing from B to A. (There are also other types of semipermeable membranes, as described in Chapter 3, p. 132.) When compartment B is filled with the solution, the liquid penetrates from A to B as long as the level of the liquid in tube T reaches a certain height. Thus, the hydrostatic pressure has increased in B, thereby preventing further penetration of the solvent from A. When we increase the solute

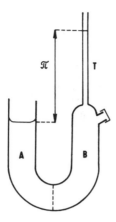

Fig. 14. A simple instrument for the study of osmotic pressure (π). Compartment A contains pure solvent, compartment B the solution

concentration in B, the final height of the solvent column in T increases in direct proportionality to the concentration.

The chance for penetrating into the pores of the membrane is greater for the molecules of the pure solvent than for the solvent molecules from the solution. In proportion to their number, the solute molecules hinder the solvent molecules from penetrating into the pores of the membrane so that there is a net flow of the solvent molecules from A to B. This phenomenon is called *osmosis*. When the hydrostatic pressure on the solution side is increased, the rate of transfer of the solvent molecules from B to A increases and counterbalances the flow in the opposite direction. This phenomenon evokes the impression that some kind of negative pressure existing in the solution draws the solvent molecules from the pure solvent. In equilibrium this *osmotic pressure* is counterbalanced by the excess hydrostatic pressure produced.

In dilute solutions the osmotic pressure π is described by the equation

$$\pi = RTx_2 \tag{1.7}$$

which resembles the Ideal Gas Law.

Accurately, the osmotic pressure depends on the content of the solvent in the solution according to the equation

$$\pi = -RT\ln x_1 = -RT\ln(1 - x_2)$$

The minus sign in this equation appears because the osmotic pressure increases with the relative decrease in the content of the solvent in the solution. When the solution is dilute ($x_2 \ll 1$) the approximate relation $\ln(1 - x_2) \approx -x_2$ can be used to yield equation (1.7).

Osmotic pressure was discovered by the botanist W. Pfeffer in 1887. It is of enormous importance for cellular processes. The distribution of water in an organism depends on osmotic pressure equilibrium in the cells and in the intercellular liquid. If a cell is in a medium of lower osmotic pressure than that inside the cell, water present in the intercellular liquid rapidly penetrates into the cell, thus increasing its volume. In an extreme case the cell will then burst. On the contrary, if the cell is in a medium of higher osmotic pressure than that within the cell, water flows out of the cell, which then shrinks. These phenomena are termed *plasmolysis* (see Fig. 15). The changes in the osmotic pressure make possible, for example, the mechanical action of tissues against their surroundings. Thus, the pressure exerted by a growing plant on the neighbouring soil is of this origin. All these phenomena depend on the specific properties of cell membranes as described in Chapter 3.

Returning to the Raoult Law, consider a solution with a very high concentration of saccharose. In this range of concentrations the vapour pressure increases more slowly with increasing concentration than would correspond to equation (1.6), as if the saccharose had a lower concentration than that actually present or as if it were less 'active'. In fact, at higher concentrations of saccharose its molecules interact through dipole–dipole and van der Waals forces, thereby decreasing their influence on water molecules. Solutions whose behaviour is no longer governed by the Raoult Law are called non-ideal.

Non-ideal properties are also exhibited by the osmotic pressure at higher solute concentrations, which is then lower than would correspond to equation (1.7). Consequently with respect to the osmotic pressure the solution behaves as more dilute than it actually is.

At the beginning of this century G. N. Lewis suggested that, in equations describing various equilibrium properties of solutions like the Raoult Law, the osmotic pressure equation and other relationships which will be considered below, concentration quantities like the molar concentration, molality, molar

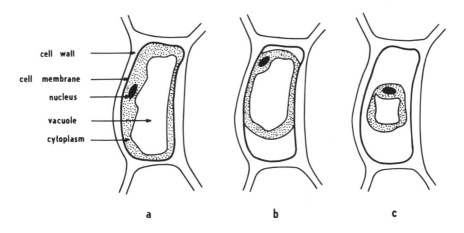

cell wall
cell membrane
nucleus
vacuole
cytoplasm

a   b   c

Fig. 15. Plasmolysis. Under normal conditions the cytoplasm spreads to the cell wall which is in contact with the total surface of the cell membrane (a). In a hypertonic solution (with an osmotic pressure larger than that inside the cell) water flows out of the vacuole and from the cytoplasm and the cel membrane separates from the cell wall (b). Scheme (c) illustrates the conditions in a strongly hypertonic solution

fraction, etc. should be replaced by the *activities*. Activity a is given by the simple relationship

$$a = \gamma c$$

where γ is the activity coefficient and c the concentration. The activity coefficient is a correction factor for non-ideal behaviour. In dilute solutions it approaches unity and the activity can then be identified with concentration. At higher concentrations it is a function of the concentration. On introducing activities, the equilibrium relationship of the Raoult Law type can be used for arbitrary concentrations. The general form of the Raoult Law is

$$(p_0 - p_1)/p_0 = a_x = \gamma_x x_2$$

In this equation a_x is the activity corresponding to the molar fraction scale and γ_x is the activity coefficient on the same scale (called the rational activity coefficient). If the content of the solute is expressed by the molar concentration, the molar activity coefficient γ_c and, for the molality scale, the molal activity coefficient γ_m are used. At the same content of the dissolved substance, different types of activity coefficient have different values, but for a dilute solution all of them approach unity.[2]

As a correction factor for the osmotic pressure, the osmotic coefficient φ is introduced, defined by the equation

$$\varphi = \pi/\pi^*$$

where π is the osmotic pressure of a real solution at a given concentration and π^* is the osmotic pressure of a solution of the same concentration which would behave ideally (i.e. according to equation (1.7)).

The situation is quite different when an electrolyte, e.g. sodium chloride, is dissolved in water. Again the water vapour pressure drops, but the decrease is larger than for a non-electrolyte. For NaCl concentrations lower than 10^{-3} mol.l^{-1} (the mole fraction is smaller than 1.8×10^{-5}) the relationship

$$(p_0 - p_1)/p_0 = 2x_2$$

is obtained.

According to this equation the solution seems to be twice as concentrated as a non-electrolyte solution of the same concentration. Obviously, on dissolution the sodium chloride completely decomposed or, better, dissociated to chloride and sodium ions. A substance that completely dissociates to ions in a particular solvent is termed a *strong electrolyte*. When the concentration of a strong electrolyte in water is increased over the concentration limit of 10^{-3} mol.l^{-1}, the decrease in the water vapour pressure again becomes slower since the ions are less active in the solution than would correspond to their concentration. Why does this happen at considerably lower concentrations than in non-electrolyte solutions? The answer can be found in the electrostatic forces acting between the ions, which act at much larger distances than short-range van der Waals forces. The influence of non-ideal behaviour is again expressed by the activity coefficients. Direct measurement does not yield the activity coefficients of the individual kinds of ions because in solution anions and cations are always simultaneously present in order to preserve the electroneutrality condition. Only the activity coefficient of the whole electrolyte, which is termed the mean activity coefficient $\gamma_\pm$, can be found. This limitation need not deter us from attempting to find the activities of individual ions, which are useful for various practical tasks. For this purpose certain assumptions must be made, as shown in the section on the hydrogen ion activity on p. 35.

Now a more general case can be considered when an electrolyte whose composition is described by a certain chemical formula dissociates in solution to n ions corresponding to that formula. For example, the dissociation of lanthanum chloride $LaCl_3$ is expressed by the equation

$$LaCl_3 = La^{3+} + 3\,Cl^-$$

giving $n = 4$. The general form of equation (1.6) for an arbitrary electrolyte composition is

$$(p_0 - p_1)/p_0 = n\gamma_{x,\pm}x_2 \tag{1.8}$$

Here $\gamma_{x,\pm}$ is the mean rational activity coefficient (i.e. for the activity given by the product of the activity coefficient and the mole fraction). The mean molar and the mean molal activity coefficients are often used as well. All three kinds of activity coefficients are related to the corresponding concentration scales.

The activity coefficients are rather important quantities for they make it possible to describe equilibria in chemical reactions in a fully quantitative way. They appear in expressions characterizing the acidity of solutions, and knowledge of activity coefficients is necessary for the analysis of solutions by potentiometry. Their values for ions present in nerve cells, for example, are important for describing the behaviour of these cells, etc. Consequently, these seemingly rather abstract quantities will be considered in some detail.

The electrostatic interaction between ions is stronger the higher the ionic charge. When a solution contains, at the same concentration, types of ions with a higher charge, they will exert a greater electrostatic effectiveness, 'strength', on the components of the solution. This influence was quantitatively expressed by G. N. Lewis as the 'ionic strength' I given by the equation

$$I = \tfrac{1}{2}(z_1^2 c_1 + z_2^2 c_2 + z_3^2 c_3 + \cdots) \tag{1.9}$$

where $z_1, z_2, z_3, \ldots$ are the charge numbers of the individual types of ions present at concentrations $c_1, c_2, c_3, \ldots$. For rather low electrolyte concentrations, Lewis found the empirical relationship

$$-\log\gamma_\pm \sim \sqrt{I}$$

The electrolyte solution is only outwardly electroneutral. Within the solution the positively and negatively charged ions move randomly in all directions depending on their thermal energy. At the same time, ions with opposite charges are attracted and those with the same charge are repulsed. Consequently, it can be expected that anions will be found in the neighbourhood of cations, and vice versa. In this way a slight surplus of charge of opposite sign, a space charge, appears in the region around every ion in the electrolyte solution. P. Debye and E. Hückel termed this domain of space charge the *ionic atmosphere*, which is not a static structure but a statistical phenomenon. There is a somewhat greater probability for an ion of opposite charge to appear for a moment in the neighbourhood of the 'central' ion than for an ion with a charge of the same sign. The region of a deviation from electroneutrality extends to only very small distances from the ion (see Fig. 16). The thickness of this region is characterized by the Debye length

$$\chi = \sqrt{\frac{\varepsilon RT}{2FI}} \qquad (1.10)$$

which obviously decreases with increasing ionic strength I or electrolyte concentration. In equation (1.10) ε denotes the permittivity of the solvent ($\varepsilon = \varepsilon_0 D$, where D is the relative permittivity and ε_0 the permittivity of a vacuum). R is the gas constant appearing, for example, in the equation of state of an ideal gas, T is the absolute temperature, and F is the Faraday constant. This quantity, which will frequently be encountered, corresponds to the charge of one mol of positively charged univalent ions and therefore has the unit $C.mol^{-1}$.

The ionic atmosphere acts electrostatically on the central ion, thus decreasing its activity. This electrostatic influence increases when the ionic atmosphere becomes more compact. Consequently, in more concentrated solutions the activity coefficient decreases (this is not true for some very concentrated solutions). For the dependence of the activity coefficient on the ionic strength in

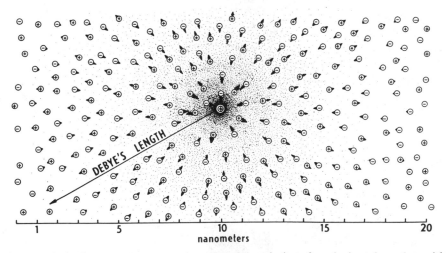

Fig. 16. The ionic atmosphere around a cation (in a solution of a univalent electrolyte with $c = 10^{-3}\ mol.dm^{-3}$). Arrows of different size indicate electrostatic interactions with the cation. The shaded area indicates the presence of space charge

rather dilute solutions, Debye and Hückel deduced, using a statistical procedure, the formula for a univalent ion k

$$\log \gamma_k = AI^{1/2} \tag{1.11}$$

and for the mean activity coefficient of a uni-univalent electrolyte (consisting of univalent cations and anions)

$$\log \gamma_\pm = -AI^{1/2} \tag{1.12}$$

Equation (1.11) is valid for the activity coefficient of an individual ion because it was obtained by a non-thermodynamic approach. The mean activity coefficient of a uni-univalent electrolyte is given by the relationship

$$\gamma_\pm = \sqrt{\gamma_B \gamma_A}$$

where γ_B and γ_A are the activity coefficients of the cation and anion, respectively. A is a function of temperature and of several basic physical constants; at 25°C, $A = 0.511 \, l^{1/2} mol^{1/2}$. The dependence of the activity coefficient on the ionic strength is depicted in Fig. 17. The actual activity coefficient deviates at rather low values of the ionic strength from the dependence described by equations (1.11) and (1.12). These simple relationships, usually called the Debye–Hückel Limiting Law, correspond to a grossly simplified picture of a solution of a strong electrolyte. The ions are represented by point charges (the influence of the size of the ion is neglected)—each of them can approach the other to an arbitrary distance and exclusively electrostatic forces act among them. It is necessary, however, to stress that the Debye–Hückel Limiting Law exactly describes a simple model of a natural situation—perhaps in the same way as the equation of state of an ideal gas perfectly describes the behaviour of an extremely diluted gas.

When the electrolyte becomes more concentrated, the ions interact in various ways. They are, of course, not point charges but have characteristic dimensions. An individual ion can only penetrate to the neighbourhood of another ion to a distance given by the sum of their radii. The ions in solution are not completely

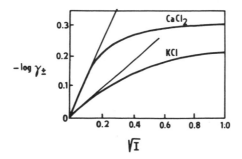

Fig. 17. Dependence of the logarithm of the mean activity coefficient of $CaCl_2$ and KCl on the square root of the ionic strength I (mol.dm^{-3}). Straight lines indicate the dependence according to the Debye–Hückel Limiting Law. From E. Hála and A. Reiser. With permission of Academia, Prague

'free' but are solvated. The interaction of solvated ions is not exclusively electrostatic but, particularly at higher concentration, short-distance forces play a role. At still higher concentrations, the ions compete for solvent molecules. In various solvents *ion association*[3] occurs to a different extent. As shown by N. Bjerrum, under definite circumstances two ions of opposite charge can more advantageously exist as an electroneutral aggregate bound together by electrostatic forces than as two separate ions in solution. Such an ion-pair is formed more easily at higher concentrations of the electrolyte and at lower permittivity of the solvent. In water $(D = 78)$ ion-pairs appear at high (at least one molar) electrolyte concentrations. However, an exact concentration limit for ion-pair formation in water cannot be determined because the effects of ion interaction are superimposed on the influence of the association. Thus, there are no adequate means of separating the individual influences. In solvents with a lower permittivity, such as nitrobenzene, acetonitrile, or methanol (D around 30), ion association is marked at concentrations of the order of $10^{-2}\,\text{mol.dm}^{-3}$. Finally, in dioxane $(D = 4)$ or in benzene $(D = 2)$ the ions are markedly associated at quite low concentrations (see Fig. 18). In addition to simple anion–cation associates, even more complicated structures like two cations–one anion, etc. can be present in these solvents.

Ion association also depends, of course, on factors other than mere interaction between ion changes. A definite role is played by ion–dipole and van der Waals forces, both of which depend on the structure of the ions.

It is difficult to develop a complete theory of strong electrolytes which would consider all the factors mentioned. The present approach, based on the methods

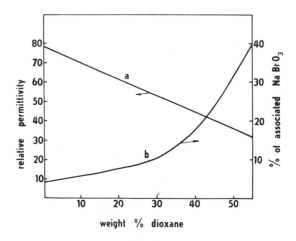

Fig. 18. Association of 0.09 M $NaBrO_3$ in dioxane–water mixtures. Water is arbitrarily miscible with dioxane so that mixtures of relative permittivity from 78.6 (pure water at 25°C) to 2.2 (pure dioxane at 25°C) can be prepared (curve a). Curve b shows the increase in the association of $NaBrO_3$ with decreasing permittivity of the solvent. According to C. A. Kraus

of statistical mechanics, first considers pair interactions (i.e. the values of the energy necessary for the approach of two ions at varying concentration and solution conditions).

Equation (1.11) is a more general expression of the activity coefficient of an individual ion than is obvious from the preceding considerations. Let us determine, for example, the activity coefficient of the zinc ion in a solution of 10^{-5} M $Zn(NO_3)_2$ and 10^{-3} M KNO_3. (The symbol M KNO_3 will denote a 'molar solution of KNO_3'.) The ions derived from the potassium nitrate do not react with the zinc ion and only exhibit an influence on its activity coefficient which depends on the total ionic strength alone. Such an excess electrolyte is termed *indifferent*. For the calculation of the activity coefficient of the zinc ion, equation (1.11) can then be applied directly.

Indifferent electrolytes are important for the study of some equilibria of reactions between ions, particularly in the field of coordination compounds (complexes).* The equilibrium between the metallic ion M, the complexing agent or ligand X, and the complex depends on the activity of the participating particles, of the reactants M and X, and of the complex as the reaction product. As a simple example consider the complexation of the cadmium ion with the anion of ethylenediaminotetraacetic acid (EDTA), which is an important reagent in

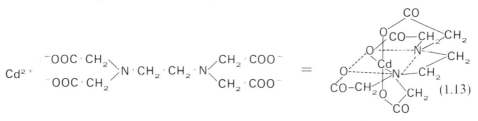

analytical chemistry. The EDTA anion will be denoted as X^{4-}. A simple complex with a metal:ligand ratio of 1:1 is produced by the reaction

$$Cd^{2+} + \ldots = \ldots \quad (1.13)$$

The equilibrium in this reaction is characterized by the stability constant of the complex

$$K_{MX} = \frac{a_{MX}}{a_M a_X} = \frac{\gamma_{MX} \cdot c_{MX}}{\gamma_M \cdot \gamma_X c_M c_X} \quad (1.14)$$

In this equation the quantities denoted a are the ion activities, the quantities denoted γ are the activity coefficients, and the quantities denoted c are the concentrations. When an excess of an indifferent electrolyte is present in the solution, the activity coefficients are virtually functions of the concentration of

---

* Equilibrium in a reaction is the situation when the reaction rate in the direction from the reactants to the products is the same as the rate in the opposite direction, i.e. from the products to the reactants (cf. p. 93 and also Appendix B).

the indifferent electrolyte alone and not of the ions participating in the equilibrium (1.13). For example, in a solution of 10^{-4} M Na_2CdX and 1 M KCl, the upper limit for the ionic strength is given by the equation

$$I = \tfrac{1}{2}(2c_{Na^+} + 2^2 c_{Cd^{2+}} + 4^2 c_{X^{4-}} + c_{K^+} + c_{Cl^-})$$

assuming complete dissociation of the complex (cf. equation (1.9)). The sum of the terms in this equation corresponding to Na_2CdX is 2.2×10^{-3} mol.dm^{-3} while the sum of those corresponding to the indifferent electrolyte KCl is 2 mol.dm^{-3}. Thus, we can neglect all the terms except c_{K^+} and c_{Cl^-}. The equation can be rewritten into the form

$$\frac{c_{MX}}{c_M \cdot c_X} = K'_{MX} = K_{MX} \cdot \frac{\gamma_M \cdot \gamma_X}{\gamma_{MX}}$$

Quantity K'_{MX} is termed the apparent stability constant. It does not vary when the indifferent electrolyte concentration is not changed and when the concentration of ions participating in equilibrium (1.13) is much lower than the concentration of the indifferent electrolyte. Then the term $\gamma_M \cdot \gamma_X / \gamma_{MX}$ is an exclusive function of the indifferent electrolyte concentration.

The compound Na_2CdX is an example of a weak electrolyte which, on dissolution in water, decomposes (dissociates) only partially to ions. The extent of the dissociation depends on the equilibrium constant which will be apparent from the more detailed discussion of acids and bases.

References

1. See the textbooks on thermodynamics, e.g. I. M. Klotz, *Chemical Thermodynamics*, Benjamin, New York, 1964.
2. See, for example, J. Koryta, J. Dvořák, and V. Boháčková, *Electrochemistry*, Chapman and Hall, London, 1973.
3. C. W. Davies, *Ion Association*, Butterworths, London, 1962.

Acidity of Solutions

Lemon juice has a sour taste. The taste of 0.1 % hydrochloric acid (the more concentrated acid would cause burns) is stronger (approximately the same as that of stomach juices which contain about 0.1 % hydrochloric acid). A glycine (aminoacetic acid) (Greek *glykys* = sweet) solution has a sweetish taste although it is termed an acid. Estimation by tasting, though very accurate with experienced people, cannot be used as an objective measure of acidity. The appropriate methods of quantitative assessment of acidity will now be considered.

Already in the distant past a link was suspected between the acidity and the composition of acid substances. At the end of the eighteenth century the French chemist Lavoisier suggested that the acid substance was oxygen. (Lavoisier formed the name of oxygen from two Greek words, *oxys* = acid, sharp, and *gennao* = bear, produce.) However, as early as the beginning of the nineteenth century the English scientist Humphry Davy showed that the very acidic hydrogen chloride only consists of chlorine and hydrogen and that hydrogen is the substance common to all acids. Finally, at the end of the nineteenth century, Swedish chemist Svante Arrhenius defined acids as compounds that can split off hydrogen ions.

A common school demonstration is the neutralization of a solution of hydrochloric acid with sodium hydroxide as a base (i.e. the transformation of an acid into a salt). An indicator (for example, bromthymol blue) is added to the solution. By a colour change (in this case from yellow to blue) this substance indicates the equivalence point where an equal concentration of the acid and the base is present in the solution. Let us carry out this experiment in a different way. After each addition of the solution of NaOH the conductance of the solution will be measured (the method of measurement is described on p. 53). In this way the dependence shown in Fig. 19 is obtained. The conductivity of a solution is roughly proportional to the number of ions present and to their mobilities. The added solution of NaOH is very concentrated, so that the additions result in only a negligible change in the total volume of the solution. Since HCl is a strong electrolyte, neutralization results in loss of a strongly conductive component from the solution, which is replaced by much less conductive ions (Fig. 19, curve 1). As there are further supporting facts for the view that the hydrogen ion is the most mobile of all ions, the neutralization process can be identified with the reaction

$$H^+ + OH^- = H_2O$$

Thus, in the solution the hydrogen ions are replaced by much less mobile sodium ions coming from NaOH and the conductance decreases. The increase of conductance following the equivalence point is caused by sodium and hydroxide ions from excess NaOH.

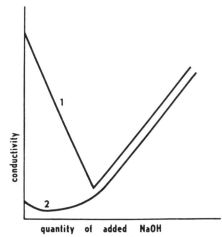

Fig. 19. Dependence of the conductivity of a solution on the degree of neutralization of a strong (1) and a weak (2) acid (conductometric titration curve)

Another situation occurs with neutralization of acetic acid (Fig. 19, curve 2). The initial increase of conductance corresponds to the neutralization of the hydrogen ions formed by dissociation of the acid in the original solution. There are, however, only a few hydrogen ions in the solution and, when they have been consumed, the conductance starts to increase again as the undissociated acid reacts with the hydroxide ions

$$CH_3COOH + OH^- = CH_3COO^- + H_2O$$

Acetic acid differs from hydrochloric acid in the simultaneous presence of undissociated CH_3COOH molecules together with acetate and hydrogen ions. Hydrochloric acid, which is completely split into hydrogen and chloride ions, is called a *strong acid*. In contrast, acetic acid, which is, in part, present in the solution in an undissociated form, is a *weak acid*. It will be shown below that the strength of acids depends on the medium, i.e. on the solvent.

The hydrogen ion cannot be considered as only a proton since it has a structure which depends on the solvent in which the acid dissociates. In water this structure is H_3O^+, the hydronium ion; in pure 'glacial' acetic acid it is $CH_3COOH_2^+$; in concentrated sulphuric acid, $H_3SO_4^+$, etc. The basic rules characterizing the acidity of solutions were formulated in 1913 by the Danish chemist J. N. Brønsted. The acid as well as the basic properties become apparent when the substance reacts with the solvent and depend on it. Substances that split-off protons in a solution are acids. Thus, for example, hydrogen chloride or acetic acid react with water according to the equations

$$HCl + H_2O = H_3O^+ + Cl^-$$
$$CH_3COOH + H_2O = H_3O^+ + CH_3COO^-$$

By an analogous experiment to that with acetic acid, we find that ammonia behaves as a base in aqueous solutions and accepts a proton from the solvent

$$NH_3 + H_2O = NH_4^+ + OH^-$$

However, in concentrated sulphuric acid, acetic acid acts as a base, i.e.

$$CH_3COOH + H_2SO_4 = CH_3COOH_2^+ + HSO_4^-$$

In water, aniline is a base

$$C_6H_5NH_2 + H_2O = C_6H_5NH_3^+ + OH^-$$

while in liquid ammonia it behaves as an acid

$$C_6H_5NH_2 + NH_3 = C_6H_5NH^- + NH_4^+$$

If the reaction of an acid HA with a solvent HS proceeds in a not very dilute solution in such a way that almost all the dissolved substance dissociates or, in other words, the equilibrium

$$HA + HS = A^- + H_2S^+ \qquad (1.15a)$$

is considerably shifted to the right-hand side, this substance is termed a *strong acid*. If, on the contrary, the resulting concentrations of the reaction products A^- and H_2S^+ is much lower than that of the original substance HA, we call it a *weak acid*. In an analogous way a base reacts with the solvent according to

$$B + HS = BH^+ + S^- \qquad (1.15b)$$

With respect to the extent to which this equilibrium is shifted to the right-hand side, the base is termed strong or weak. Obviously in the case of a base no dissociation takes place but a transformation from an uncharged form B to the charged form BH^+ occurs. Therefore it is more appropriate to use the term *ionization* for acid–base equilibria.

Evidently, the anion of an acid acts as a base according to the equation

$$A^- + HS = HA + S^-$$

Therefore the anion A^- is termed the conjugate base to the acid HA, in the same way as BH^+ is the conjugate acid to the base B. The *protolytic* reaction is the reaction of an acid with a solvent or with another basic species to form a conjugate base.

Acids which split-off several protons (oligoprotonic acids, Greek *oligoi* = several) such as, for example, orthophosphoric acid H_3PO_4, undergo protolytic reactions to give species which can act as acids or bases at the same time. Thus, the $H_2PO_4^-$ anion reacts in water as an acid

$$H_2PO_4^- + H_2O = HPO_4^{2-} + H_3O^+$$

or as a base

$$H_2PO_4^- + H_2O = H_3PO_4 + OH^-$$

The strength of the acid is expressed by the equilibrium constant of reaction (1.15a)

$$K = \frac{a_{A^-} \cdot a_{H_2S^+}}{a_{HA} \cdot a_{HS}} \qquad (1.16)$$

In a dilute solution the solvent activity a_{HS} can be considered constant and included in the equilibrium constant in equation (1.17)

$$K_{d,S} = K \cdot a_{HS} = \frac{a_{A^-} \cdot a_{H_2S^+}}{a_{HA}} \qquad (1.17)$$

For an aqueous solution

$$K_d = \frac{a_{A^-} \cdot a_{H_3O^+}}{a_{HA}} \qquad (1.18)$$

The larger K_d, the greater is the dissociation and the stronger is the acid. For acids that are strong electrolytes, the dissociation is complete and the equilibrium constant meaningless. This class of acids includes perchloric, chloric, hydrochloric, nitric, sulphuric acids (for the first degree of dissociation H_2SO_4

$+ H_2O = HSO_4^- + H_3O^+$), etc., of course, in aqueous solutions. The strength of an acid can be advantageously characterized by the logarithm of the dissociation constant multiplied by -1. This quantity is denoted by the symbol

$$-\log K_d = pK_d \qquad (1.19)$$

Obviously, the larger pK_d, the weaker is the acid. Table 3 lists the pK_d-values of more important weak acids. The relations for the ionization constants of bases can be expressed in a similar way using equation (1.15b). There is, however, an advantage in using equation (1.15b) in the direction from right to left, and in characterizing the ionization of a base by the dissociation constant of the conjugate acid BH^+, which yields, for example, for ammonia

$$K_d = \frac{a_{NH_3} \cdot a_{H_3O^+}}{a_{NH_4^+}} \qquad (1.20)$$

The higher the pH, the stronger is the base. Examples of pK_d-values for more important bases are also listed in Table 3.

When the conductivity of water purified by multiple distillation in a quartz or platinum still is measured, we find that, in spite of maximum control of the purity, a definite small conductivity remains. This conductivity is due to the self-ionization of water

$$2 H_2O = H_3O^+ + OH^-$$

The equilibrium constant defined in the same way as in equation (1.18) is called the ion product of water

$$K_W = K \cdot a_{H_2O}^2 = a_{H_3O^+} \cdot a_{OH^-} = 10^{-14} \, mol^2.dm^{-6} \qquad (1.21)$$

at 25°C.

Table 3   Dissociation constants of weak acids and weak base cations (only the name of the base, not the cation, is indicated)

Acid	pK_d	°C	Base	pK_d	°C
Formic	3.73	20	Ammonia	9.23	20
Acetic	4.76	25	Methylamine	10.64	25
Monochloro					
acetic	2.86	20	Ethylamine	10.67	25
Dichloro			Trimethyl		
acetic	1.29	25	amine	9.76	20
Phosphoric			Pyridine	5.19	25
—H_3PO_4	2.13	20			
Phosphoric					
—$H_2PO_4^-$	7.21	20			
Boric	9.28	20			

In an analogous way, the self-ionization occurs in methanol, acetic acid, liquid ammonia, and many other solvents according to the general equation

$$2\,HS = H_2S^+ + S^-$$

These solvents are called *protic* since the self-ionization proceeds by proton transfer from one solvent molecule to another. *Aprotic* solvents contain either no hydrogen atoms at all, like liquid sulphur dioxide, or these atoms are so strongly bound that proton transfer is not feasible. This group includes important solvents like benzene, dioxane

$$O\underset{CH_2-CH_2}{\overset{CH_2-CH_2}{<\;\;\;>}}O$$

dimethylsulphoxide $(CH_3)_2SO$, acetonitrile CH_3CN, dimethylformamide $HCO.N(CH_3)_2$ and hexamethylphosphortriamide $[(CH_3)_2N]_3PO$.

In protic solvents the proton of an acid is bound to a solvent molecule with different strength. Therefore the equilibrium of equation (1.15a) is shifted to a varying extent to the right-hand side, depending on the nature of the acid and of the solvent. As already mentioned, different substances act as acids of varying strength, or even as bases, depending on the character of the solvent. Water, an amphiprotic solvent, exhibits both acid and basic properties. Solvents like ammonia that are bases in aqueous solution will readily split-off protons from the acids and, in this way, increase their strength but, on the contrary, will decrease the strength of the bases (in comparison with water). Solvents of this sort are termed *protophilic*.

On the other hand, solvents that act as acids in aqueous solutions (for example, acetic acid) tend to weaken acids and strengthen bases. They are called *protogenic* solvents.

As mentioned on p. 28 the low permittivity of a solvent encourages ion-pair formation. In such solvents all acids and all bases are weak because full ionization is impossible. Because of their low permittivity, aprotic solvents like benzene or dioxane are unable to solvate protons. No protolytic reaction can occur in such media. This is, of course, not true for all aprotic solvents because solvents with high permittivity like dimethylformamide, dimethylsulphoxide, or acetonitrile are suitable media for protolytic reactions of acids. They are called *basic aprotic solvents*.

Activity of Hydrogen Ions

The acidity of aqueous solutions[1] is a result of the presence of the hydronium ion H_3O^+ (it is usual to call it the hydrogen ion). The solvation of protons in water is not completed by formation of H_3O^+ ions because more complex structures are formed, the most stable being $H_9O_4^+$

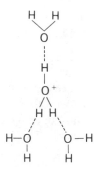

The concentration, or, more accurately, the activity, of hydronium ions, decisively determines the course of very many chemical reactions and is of profound importance for processes in living organisms. The hydrogen ion concentration in blood plasma is 4×10^{-8} mol.dm^{-3}, while in stomach juices it reaches a value of 0.03 mol.dm^{-3}. The growth of some organisms is possible in a very narrow range of hydronium activities, while the reproduction of others (moulds, for example) is only prevented by either extremely high or extremely low hydronium ion concentrations.

In aqueous solutions the interval of the values of hydronium activities is very broad. In 10^{-3} M HCl it approaches a value of 10^{-3} mol.dm^{-3} (in fact, it is somewhat lower since the activity coefficient is 0.97). The value of the hydronium activity in 10^{-3} M NaOH can be calculated using the ion product of water (equation 1.21)). The hydroxide ion activity in this solution is 0.97×10^{-3} mol.dm^{-3} and equation (1.21) yields $a_{H_3O^+} = 1.03 \times 10^{-11}$ mol.dm^{-3}. In view of this enormous range of hydronium ion activities it is preferable not to give the activities themselves but their logarithmus multiplied by -1.

In textbooks on thermodynamics the activity is usually defined as a dimensionless number, the dimension of the activity coefficient being the reciprocal of the appropriate concentration quantity. In a book destinated for non-specialists in physical chemistry, it seems more advantageous to consider the activity as a kind of concentration corrected for various interactions in the system under investigation. The main difficulty in this treatment arises from the logarithmic functions of the activity, because mathematicians require dimensionless quantities as variables in logarithmic functions. We shall circumvent this inconsistency by tacit assumption that in functions like log a the activity a (with units of mol.dm^{-3}, for example) is, in fact, divided by unit activity so that the operator log is applied to a dimensionless quantity.

This quantity is denoted as pH

$$pH = -\log a_{H_3O^+} \tag{1.22}$$

This simple definition is connected with a certain complication inherent in the fact that $a_{H_3O^+}$ is the activity of an individual sort of ion. However, this question will not be considered in detail, and it will simply be assumed that the activities of hydronium ions exist and that it is completely satisfactory to use the Debye–Hückel Limiting Law (1.11) for determining the pH of very dilute strong acids and alkali hydroxides. Furthermore, the pH-values of certain solutions have been standardized (see Table 4) and the hydronium activities of unknown solutions are determined by comparison with these standards (see p. 137).

The pH of a 2.7×10^{-4} M HCl solution is 3.73. When sodium hydroxide is added to the solution so that its resulting overall concentration is 3.7×10^{-4} M the pH of the solution reaches a value of 10. The sodium hydroxide has neutralized the hydronium ions coming from the hydrochloric acid and the remaining concentration, 10^{-4} mol.dm^{-3} of hydroxide ions, decides the final pH-value. In a second experiment, a solution containing 5×10^{-2} M acetic acid and 5×10^{-3} M sodium acetate is prepared, with a pH of 3.73. When the same amount of sodium hydroxide as in the preceding experiment is added to the solution, the pH changes only slightly to 3.77. The mixture weak acid/weak-acid salt keeps the pH close to the original value. The same influence is exhibited when hydrochloric acid is added in the same amount. Thus, this mixture acts as a sort of a buffer against the invading hydronium or hydroxide ions. In the solution of

Table 4 pH of five standard buffer solutions at different temperatures

°C	A	B	C	D	E
0		4,003	6,984	7,534	9,464
5		3,999	6,951	7,500	9,395
10		3,998	6,923	7,472	9,332
15		3,999	6,900	7,448	9,276
20		4,002	6,881	7,429	9,225
25	3,557	4,008	6,865	7,413	9,180
30	3,552	4,015	6,853	7,400	9,139
35	3,549	4,024	6,844	7,389	9,102
40	3,547	4,035	6,838	7,380	9,068
45	3,547	4,047	6,834	7,373	9,038
50	3,549	4,060	6,833	7,367	9,011
55	3,554	4,075	6,834		8,985
60	3,560	4,091	6,836		8,962
70	3,580	4,126	6,845		8,921
80	3,609	4,164	6,859		8,885
90	3,650	4,205	6,877		8,850
95	3,674	4,227	6,886		8,833

A = potassium hydrogen tartarate (sat. at 25°C).
B = potassium hydrogen phthalate (0.05 mol/kg).
C = KH_2PO_4 (0.025 mol/kg) + Na_2HPO_4 (0.025 mol/kg).
D = KH_2PO_4 (0.008695 mol/kg) + Na_2HPO_4 (0.03043 mol/kg).
E = $Na_2B_4O_7$ (0.01 mol/kg).

From R. G. Bates, *Determination of pH, Theory and Practice*, 2nd Edition, John Wiley & Sons, New York, 1973.

5×10^{-2} M acetic acid alone the acid is dissociated only to a small degree. Again when sodium acetate alone is present in the solution, it completely dissociates to sodium cations and acetate anions. These anions further *hydrolyse* to a small degree, i.e. they combine with water protons. Some undissociated acetic acid is formed and hydroxide ions are set free

$$CH_3COO^- + H_2O = CH_3COOH + OH^- \tag{1.23}$$

Therefore a solution of a weak acid (CH_3COOH) with a strong base ($NaOH$) is alkalinous. When excess acetic acid is present, it suppresses the hydrolysis (1.23) and shifts the equilibrium to the left-hand side. On the other hand, the excess acetate ions decrease the dissociation of the acetic acid. In the final result the excess acetic acid is virtually in an undissociated form, while the concentration of the excess salt is equal to the concentration of the free anion of the acid. Thus, for calculation of the *pH* of the solution equation (1.18) can be used which, in the logarithmic form, yields the Henderson–Hasselbalch equation

$$pH = pK_d + \log(a_{A^-}/a_{HA})$$

$$\approx pK_d + \log\frac{\gamma_{A^-} \times \text{salt concentration}}{\text{weak-acid concentration}} \tag{1.24}$$

This equation indicates the *pH* of a mixture of a weak acid (acidic component) and its salt with a strong base (basic component) called a *buffer*. In an analogous way a mixture of a weak base and its salt with a strong acid also acts as a buffer (for example, ammonia/ammonium chloride). The Henderson–Hasselbalch equation for this system can be deduced on the basis of equation (1.20).

If a definite amount of a strong base is added to the buffer, part of the undissociated acid is transformed to the salt. The new *pH*-value can be calculated by equation (1.24). The extent of the *pH* change is smaller, the smaller is the concentration of the newly formed salt compared with the original concentrations of the acid and the salt. The buffering capacity increases with increasing concentration of both components of the buffer. Table 5 contains some recommended buffer solutions.

The preservation of a constant hydronium activity using buffers is particularly unavoidable in the study of organic reactions, in biochemistry, and in microbiology. The buffering systems are created by the organism itself. Blood plasma maintains a *pH* between 7.35 and 7.45. It contains various species of acidic and basic character such as proteins and organic acids, but the main influence is exhibited by the buffer carbonic acid H_2CO_3/bicarbonate ion HCO_3^-. The acidic component, H_2CO_3, is the product of oxidation of a variety of carbonaceous substances in the organism. The original product, carbon dioxide, which is in equilibrium with the carbonic acid, escapes from the blood in the lungs. The transformation of carbonic acid to carbon dioxide is a comparatively slow reaction but it is catalysed by the enzyme, carbonate dehydratase. Because of the continuous removal of carbon dioxide from blood, the concentration of carbonic

Table 5   Buffers with constant ionic strength $I = 0.2$

pH	1	2	3	Basic solutions 4	5	6	7	8	9
2.0	72.0	10.6	14.7						
2.5	72.0	22.8	8.6						
3.0	72.0	31.6	4.2						
3.5	72.0	36.6	1.7						
4.0	72.0				20.0	33.7			
4.5	72.0				20.0	11.5			
5.0	72.0				20.0	3.7			
5.5	72.0				20.0	1.2			
6.0	72.0						9.2	6.6	
6.5	72.0						16.6	3.7	
7.0	72.0						22.7	1.6	
7.5	72.0						24.3	0.5	
8.0	72.0		10.4						80.0
8.5	72.0		5.3						80.0
9.0	72.0		2.0						80.0
9.5	72.0	34.5		2.7					
10.0	72.0	28.8		5.6					
10.5	72.0	23.2		8.4					
11.0	72.0	19.6		10.2					
11.5	72.0	17.6		11.2					
12.0	72.0	15.2		12.4					

The indicated volume of basic solutions (in cm^3) is made up to 200 cm^3. Basic solutions: 1, 5 M NaCl; 2, 1 M glycine; 3, 2 M HCl; 4, 2 M NaOH; 5, 2 M sodium acetate; 6, 3.5 M acetic acid; 7, 0.5 M Na_2HPO_4; 8, 4 M NaH_2PO_4; 9, 0.5 M sodium salt of veronal.

From H. M. Rauen, Ed., *Biochemisches Taschenbuch*, Vol. II, Springer-Verlag, Berlin, 1964, p. 104. Reproduced by permission of Springer-Verlag, Heidelberg

acid is considerably lower than that of the bicarbonate anion (0.027 mol.dm^{-3}), so that the resulting pH is higher than 7 (pK of carbonic acid is 6.1).

Aqueous solutions of salts of strong acids and bases (for example, KCl and Na_2SO_4) should exhibit a neutral pH (pH = 7). However, such salt solutions exposed to the air are always acidic, with pH about 5.5. This seemingly surprising finding is due to dissolution of omnipresent carbon dioxide in water with a resulting acidic reaction. The meaning of the pH in minute formations in cells, called organelles, is often questioned (the mitochondrion, chloroplast, cytoplasmatic reticulum, etc. belong in this group). For the sake of simplicity consider a spherical organelle with a diameter of 0.1 μm. Its volume is approximately $4 \times 10^{-21} m^3 = 4 \times 10^{-18} dm^3$. If the inside of the organelle has pH = 7, it only contains approximately 4×10^{-25} mol H_3O^+, i.e. only one-quarter of an ion. This result might lead to the conclusion that the pH of such small formations is meaningless. However, this view is not correct, because activity is a statistical term. By dissociation of the acid component of the buffer inside the organelle, hydronium ions are formed but subsequently disappear by reaction with the basic component of the buffer. The time average seems to indicate that a fraction of a hydronium ion is present in the solution. In fact, the acidity of the solution is determined by the acidic and the basic buffer components which are represented inside the organelle by sufficiently large numbers of particles.

Reference

1. R. G. Bates, *Determination of pH*, John Wiley & Sons, New York, 1964.

Acidic and Basic Properties of Aminoacids and Natural and Synthetic Polymers

Fatty acids substituted by amine and eventually by other groups, the aminoacids, are the building units of proteins and of simpler peptides which often have important biological functions, for example, hormones and antibiotics. The simplest aminoacid is glycine or aminoacetic acid (2-aminoethoic acid). Its formula according to this name would be $NH_2 \cdot CH_2 \cdot COOH$, which obviously has an acidic group ($-COOH$) as well as a basic group ($-NH_2$) in its molecule. Such substances are called ampholytes. It is then of interest to find the conditions under which the acidic or the basic function becomes decisive. In a strongly acidic medium glycine is present in the form $NH_3^+ \cdot CH_2 \cdot COOH$. From the measurement of the ionization of glycine it follows that at a *pH* of approximately 2, one hydrogen ion is split-off. The resulting particle has a high dipole moment which can be explained solely by the presence of ionic charges in the aminoacid molecule. A particle with the structure $NH_2 \cdot CH_2 \cdot COOH$ could not have such a high dipole moment. Furthermore, we cannot expect that the substitution of the $CH_2 \cdot COOH$ group into the ammonia molecule NH_3, which is a base of medium strength ($pK = 9.2$), would lead to the formation of an extremely weak base with $pK \approx 2$. The only explanation of these two facts is based on the assumption that the protolytic reaction

$$NH_3^+ \cdot CH_2COOH + H_2O = NH_3^+ \cdot CH_2 \cdot COO^- + H_3O^+$$

takes place at *pH* 2. The ion $NH_3^+ \cdot CH_2.COO^-$ is called an amphion or zwitterion (German *zwitter* = hermaphroditic). This ionic form is stable in aqueous solutions and the tautomeric reaction*

$$NH_3^+ \cdot CH_2 \cdot COO^- = NH_2 \cdot CH_2 \cdot COOH$$

almost does not take place. At *pH* as high as 9 a further protolytic reaction ensues

$$NH_3^+ \cdot CH_2 \cdot COO^- + H_2O = NH_2 \cdot CH_2 \cdot COO^- + H_3O^+$$

* Tautomerization (Greek *tautos* = identical, *meros* = part) is a reaction in which the molecules of the reactant and of the reaction product have the same number and sort of atoms but different structure.

Thus, the carboxylic group of glycine is more acidic than that of acetic acid because of the repulsive action of the positively charged NH_3^+ group on the hydrogen ion present in the carboxylic group. The $-NH_2$ group is more basic in glycine than in ammonia because the carboxylate group $-COO^-$ electrostatically stabilizes the hydrogen ion in the ammonium group $-NH_3^+$.

Consequently, the zwitterion $NH_3^+ \cdot CH_2 \cdot COO^-$ with a positive as well as a negative charge is the completely prevailing form of glycine in the pH-range between 4 and 8. This range is termed isoelectric (Greek *isos* = equal). Analogous behaviour to that of glycine is exhibited by all aminoacids which contain a single carboxylic and a single amino group as acidic or basic functional groups. A slightly different situation arises with histidine

$$
\begin{array}{c}
\text{N}-\text{C}-\text{CH}_2-\text{CH}-\text{COO}^- \\
\| \quad \| \qquad\qquad | \\
\text{CH} \ \text{CH} \qquad\quad \text{NH}_3{}^+ \\
\diagdown_{\text{N}}\diagup \\
\text{H}
\end{array}
$$

which contains the weakly basic imidazolyl group ($pK = 5.8$). The pK of the ammonium group is 9.2. In the pH range where the amphion is the prevailing form of histidine, the presence of positively charged ions

$$
\begin{array}{c}
\text{N}-\text{C}-\text{CH}_2-\text{CH}-\text{COO}^- \\
\| \quad \| \qquad\qquad | \\
\text{CH} \ \text{CH} \qquad\quad \text{NH}_3{}^+ \\
\diagdown_{\overset{+}{\text{N}}}\diagup \\
\text{H} \quad\ \text{H}
\end{array}
$$

or negatively charged ions

$$
\begin{array}{c}
\text{N}-\text{C}-\text{CH}_2-\text{CH}-\text{COO}^- \\
\| \quad \| \qquad\qquad | \\
\text{CH} \ \text{CH} \qquad\quad \text{NH}_2 \\
\diagdown_{\text{N}}\diagup \\
\text{H}
\end{array}
$$

cannot be neglected. Thus, the isoelectric pH-range shrinks to an *isoelectric point* (pH_{iso}), where the concentration of positively charged ions is equal to that of negatively charged ions. From a simple calculation it follows that

$$pH_{iso} = \tfrac{1}{2}(pK_2 + pK_3)$$

where pK_2 and pK_3 are the dissociation constants of the ammonium group of the imidazolyl and of NH_3^+. In the isoelectric pH-range or in the neighbourhood of the isoelectric point the solubility of the aminoacids is lowest and they do not exhibit any mobility in an electric field.

Peptides are formed from aminoacids by the peptidic bond

$$
\begin{array}{c}
\text{O} \ \ \text{H} \\
\| \quad | \\
-\text{C}-\text{N}-
\end{array}
$$

If they are derived from aminoacids that have one amino and one carboxylic group, the resulting chain includes ionizable groups only at both ends. In the peptidic chain of natural polymeric peptides, the proteins, there frequently occur aminoacids originally possessing more than two ionizable groups, like histidine, glutamic and aspartic acids with one more carboxylic group, lysine with a further amino group, arginine with a quanidine group, etc. The polymer then contains a large number of ionized or ionizable groups and is termed a *polyelectrolyte*. Polyelectrolytes also include nucleic acids which, in a neutral solution, have a negative charge on every nucleotide unit owing to the dissociation of the phosphate group (see Fig. 20). In addition, purine and pyrimidine bases contribute to the ionization equilibrium of nucleic acids in acidic solutions. Many polyelectrolytes are of a synthetic nature (for some examples see Table 6).[1]

The characteristic phenomenon connected with polyelectrolytes in which they differ from low-molecular electrolytes is the considerable electrical charge density in the space enclosed by the macromolecules (for example, in the form of an α-helix or disordered coil; see Fig. 21) and in its immediate surroundings. This space charge does not depend on the concentration of the polyelectrolyte. With respect to the electroneutrality condition in this space, there must be a concentration of 'small' ions corresponding to the number of opposite charges fixed in the structure of the polymer. Those ions are termed *counterions* or *gegenions* (German *gegen* = against). In the neighbourhood of the ionizable group those ions form a complete analogy of the ionic atmosphere (see p. 26). When these ionic clouds come close together (where there are a sufficiently high number of ionized groups in the macromolecule), they repulse each other and the polyelectrolyte chain expands. For example, this effect contributes to the distortion (denaturation) of the double helix of deoxyribonucleic acid, which occurs when its stability is decreased by increasing temperature. When a sufficient amount of low-molecular electrolyte is added to the polyelectrolyte solution, it penetrates even to regions where only ionized groups and gegenions were originally present. Consequently, a certain concentration of ions with the same sign as the charges fixed in the polymeric chain is formed (they are termed *coions*). The increased electrolyte concentration in the surroundings of the macromolecule results in a decrease in the dimensions of the ionic atmospheres and in suppressing the repulsive influence of the gegenions. For example, in this way the stability of the double helix of deoxyribonucleic acid is increased so that a considerably higher temperature is required for its denaturation than without the addition of the excess low-molecular electrolyte. The double-helical structure is further reinforced by the addition of magnesium ions which form complexes with phosphate groups and in this way decrease the total negative charge in the polyelectrolyte.

When alkali metal hydroxide is continuously added to a solution of a polyelectrolyte that exhibits weekly acidic or ampholytic properties (i.e. the substance is *titrated*), the acid groups ionize. For example, in polyacrylic acid the hydrogen ion is initially bonded with equal strength to each carboxylic group. When the macromolecule ionizes, the resulting negative charge prevents further

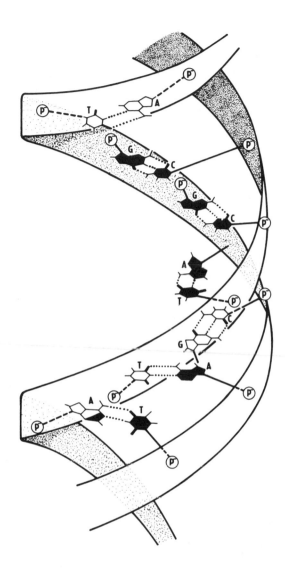

Fig. 20. The double helix of DNA. Two helical bands represent the sugar-phosphate backbone of the structure. P^- denotes the phosphoric acid unit which is bound through the ester bonds to the neighbouring molecules of the sugar, deoxyribose. The remaining OH group is ionized. The inner part of each chain consists of a sequence of four bases: adenine (A), cytosine (C), guanine (G), and thymine (T). There are two possible links of the bases from one chain to the other: either adenine is bound to thymine by two hydrogen bonds or guanine is paired to cytosine by three hydrogen bonds (dotted lines). The bases are linked to the backbone by thick lines, either full (when the bond is seen) or dashed (when the bond is hidden by the backbone). In neutral media, which are actually present in organisms, the bases are not ionized. Thus, in a solution of low ionic strength the repulsing forces between the negative charges present on the phosphate units would distort the structure, with subsequent denaturation. However, a sufficient electrolyte concentration present in the organism suppresses the electrostatic field between the anion groups

Table 6   Examples of polyelectrolytes with acidic or basic function

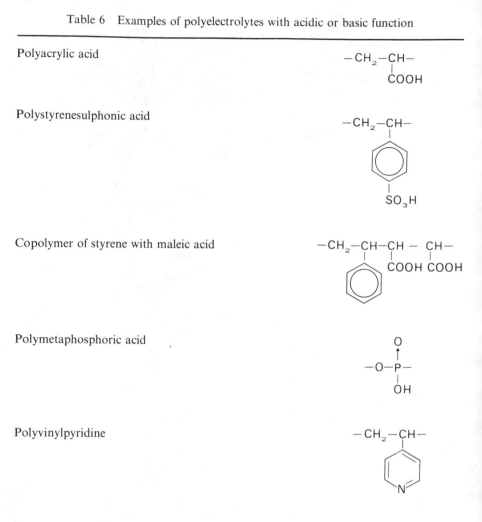

Polyacrylic acid

Polystyrenesulphonic acid

Copolymer of styrene with maleic acid

Polymetaphosphoric acid

Polyvinylpyridine

Poly-4-vinyl-N-dodecylpyridinium
(a 'polymeric soap')

Copolymer of acrylic acid with 4-vinylpyridine
(a polyampholyte)

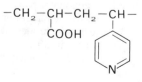

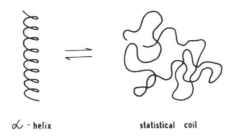

α - helix   statistical   coil

Fig. 21. Transition of an α-helix to a statistical coil

splitting-off of hydrogen ions so that the acidity of the remaining carboxylic group decreases. In this situation each carboxylic group will possess a definite effective pK-value which is higher than that of monomeric acrylic acid. This change of pK helps to elucidate the dependence of the pH of a solution, e.g. of a protein, on the amount of hydroxide added (the *titration curve*).

Strongly alkaline or strongly acidic media also have a strong impact on the behaviour of proteins. Under these conditions the acidic or basic groups in the macromolecule ionize to such an extent that an α-helix–disordered coil transition takes place (see Fig. 21). This process is again termed protein denaturation. A protein can also be denatured by elevated temperature (the helix breaks down due to excessive thermal vibration of the chain) or by the addition of urea (the hydrophobic interaction holding an ordered structure together ceases to exist). A denaturated protein loses its functions (for example, an enzyme becomes inactive) and its solubility in water decreases so that coagulation takes place.

Reference

1. H. Morawetz, *Macromolecules in Solution*, John Wiley & Sons, New York, 1965.

Ion Motion in Solution

Two glass vessels 1 and 2 containing potassium chloride solutions with two different concentrations c_1 and c_2 are connected by a thin short glass capillary (Fig. 22). The solutions in both vessels are stirred so that the concentration of KCl is constant in each vessel. However, the stirring has no effect in the capillary. The concentration changes are measured in both vessels by a suitable sensor (for example, a potassium ion-selective electrode, see p. 150, or, still better, by measurement of conductivity in vessel 2, see p. 53). There is a slight delay before a uniform drop of concentration of KCl in the capillary is established between values c_1 and c_2 and then the measurement can begin. The results show that the rate of penetration of the electrolyte, v, is directly proportional to the concentration difference between the two vessels and inversely proportional to the length of the capillary, l

$$v = k(c_1 - c_2)/l \tag{1.25}$$

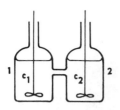

Fig. 22. A device for the study of steady-state diffusion. Cells 1 and 2 are filled with an electrolyte solution of concentrations c_1 and c_2. A constant concentration is maintained in each cell by stirring. Diffusion is restricted to the connecting capillary only

In a different electrolyte the same dependence is obtained with a different value of the proportionality constant k. The only influence acting on the motion of the ions is the concentration difference (consequently, the osmotic pressure difference); the stirring of the solutions plays no role. The process of transport of a substance controlled by the concentration difference of that substance is called *diffusion*. It is governed by the Fick Law, which can be deduced by generalization of equation (1.25). The rate of any transport process (not only diffusion) is characterized by its *flux*, which is the amount of substance (in mols) or another quantity, e.g. heat, which passes through a unit area ($1 \, m^2$ or $1 \, cm^2$) during 1 second. When the concentration changes only along a single coordinate, x (the *linear diffusion*), the Fick Law states that the flux is proportional to the concentration gradient, dc/dx

$$J = -D\frac{dc}{dx} \tag{1.26}$$

The proportionality constant D (unit $cm^2.s^{-1}$ when the unit of flux is $mol.cm^{-2}.s^{-1}$ and the unit of concentration $mol.cm^{-3}$) is termed the diffusion coefficient, or diffusivity, and specifies the diffusion rate of the given substance. The minus sign shows that the diffusion flux is oriented from higher to lower concentrations.[1]

In a physical or chemical process a characteristic quantity, y, usually changes with time (for example, the concentration of a reactant during a chemical reaction or the temperature of a body during heating) so that the differential quotient of this quantity with respect to time, dy/dt is different from zero. In certain cases, however, the process approaches a steady state, so that the rate of change of the characteristic quantity (usually a function of the coordinates of the system) approaches zero. This situation prevails in our experiment: a steady concentration distribution is established in the capillary, where the concentration decreases linearly with distance

$$\frac{dc}{dx} = \frac{c_1 - c_2}{l} \tag{1.27}$$

In other words, the same quantity of potassium chloride that came from vessel 1 into the capillary leaves it by diffusion into vessel 2.

However, many diffusion processes do not take place at steady state but have a transient nature, i.e. the concentrations are time-dependent. In the next experiment the vessel (see Fig. 23) is filled with a solution of a coloured salt, for example potassium permanganate. A long capillary containing pure water is

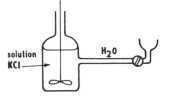

Fig. 23. A device for the study of linear diffusion in an unsteady state. The solution in the cell is stirred so that its concentration is constant in the whole cell. In the capillary the solute diffuses into pure water

attached to the vessel. The solution in the vessel is stirred so that the concentration of $KMnO_4$ is constant everywhere in the vessel, while the stirring has no effect in the capillary. Since the volume of the capillary is negligible compared with the volume of the vessel, the quantity of potassium permanganate penetrating by diffusion has virtually no influence on the overall composition of the solution in the vessel. The concentration of potassium permanganate at any distance from the orifice of the capillary is measured with a special micro-spectrophotometer. The capillary is rather long so that the penetration of potassium permanganate is observed only in the region close to the vessel. Thus, the principal features of the diffusion process in the capillary are the same as the physical model of linear diffusion in an infinitely long tube.

When the concentration of potassium permanganate is measured at different distances from the orifice as a function of time, the dependence shown in Fig. 24 is

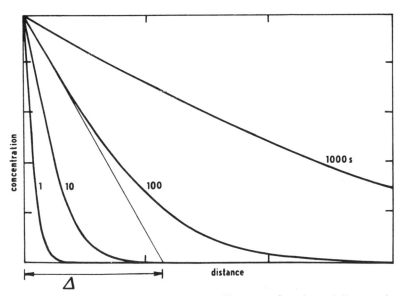

Fig. 24. Concentration distribution in the capillary as a function of distance from its orifice into the cell (cf. Fig. 22) for different values of time elapsed from the start of the diffusion. The effective diffusion layer thickness is indicated on the diagram for a time of 100 s

48

obtained. Obviously, the slope of the concentration at the origin of the coordinate system decreases with time, i.e. the diffusion rate diminishes. The concentration gradient for $x = 0$ is described by the equation (which is deduced theoretically but corresponds exactly to our experiment)

$$dc/dx = c^0/(\pi Dt)^{1/2} \qquad (1.28)$$

where c^0 is the concentration of potassium permanganate in the cell and t the time elapsed from the start of the experiment when the capillary came into contact with the solution in the vessel. In Fig. 24 we observe that the region into which $KMnO_4$ has penetrated expands with time. This region is called the *diffusion layer*. Since this layer has, of course, no definite border we must define a quantitative measure for its effective thickness. A tangent is constructed to the concentration curve at point $x = 0$. According to equation (1.28) the distance of its intersection with the line $c = 0$ (x-axis) from the coordinate origin is

$$\Delta = (\pi Dt)^{1/2}$$

Length Δ is termed the effective diffusion-layer thickness. The result of a calculation indicates that, at distance $x = \Delta$ from the orifice of the capillary, the concentration of the diffusing substance decreases to approximately 21 % of the original value. This decrease is independent of the diffusivity as well as of the concentration. The effective diffusion-layer thickness gives an illustrative picture of the rate of propagation of a substance by diffusion.

Now, the first experiment at the beginning of this section will be modified somewhat. As shown in Fig. 25, into each of vessels 1 and 2 is placed a silver–silver chloride electrode which is a silver plate covered with insoluble silver chloride. More will be said about the properties of this electrode on p. 75. Both electrodes

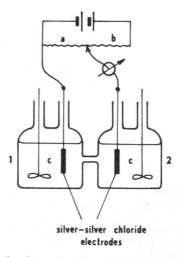

silver–silver chloride
electrodes

Fig. 25. A device for the study of ion migration. In both cells the electrolyte concentration is the same and migration occurs in the connecting capillary only

are connected to a d.c. voltage source. This source can be, for example, a potentiometer connected to a storage battery. The voltage that is fed to the electrodes from the potentiometer depends on the position of the sliding contact on the potentiometric wire (the resistance wire, usually made from an iron–nickel alloy has a constant cross-section). The resulting voltage is given by the relationship

$$U = U_{st}\,\overline{AS}/\overline{AB}$$

where $\overline{AS}$ is the length of the segment of the potentiometric wire between its origin and the position of the sliding contact, $\overline{AB}$ is the total length of the potentiometric wire, and U_{st} is the voltage of the storage battery. The electrical circuit also includes a galvanometer G which measures the electrical current flowing between the voltage source, the vessels, and the connecting capillary. Silver–silver chloride electrodes are used to obtain an electrical potential difference between the electrodes in the solution that is just equal to the voltage supplied by the potentiometer and to prevent electrode polarization (loss of voltage at electrodes under current flow, see in detail on p. 63). In contrast to the experiment described at the beginning of this section, both vessels and the capillary are filled with a potassium chloride solution of the same concentration.

The current flowing between the two compartments is directly proportional to the voltage imposed on the electrodes and to the potassium chloride concentration (at least at lower concentrations), inversely proportional to the length of the capillary, and directly proportional to its cross-section. The current increases with increasing temperature of the electrolyte. When positive charge Q has passed from vessel 1 to vessel 2, the increase in the content of potassium chloride in vessel 2 is $0.49Q/F$, while the content of KCl in vessel 1 has decreased by the same amount (the amount of KCl in the capillary is negligible). The quantity $F - 96480\,\mathrm{C/mol}$ is again the Faraday constant (p. 26), corresponding to a charge of 1 mol of univalent ions.

On the basis of these observations, it can be concluded that the flow of electric current through an electrolyte follows Ohm's Law and, therefore, the electrolyte conductivity is independent of the voltage fed to the electrodes. The system of two vessels joined by a thin capillary with a voltage source connected to the electrodes can be represented by the substitution diagram shown in Fig. 26. The resistances of the compartments, R_1 and R_2, are much smaller than the resistance of the capillary, R_c. The entire voltage of the source therefore acts between the ends of

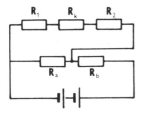

Fig. 26. A substitution diagram for the device in Fig. 25. The resistance of the solution in the capillary is much larger than that of cells 1 and 2

the capillary. Thus, the length and cross-section of the capillary determine the resistance of the whole system, as has already been observed.

The change in the amount of KCl contained in each vessel can be explained in the following way. When a positive charge is injected from the silver–silver chloride electrode into the solution in vessel 1, the reaction

$$Ag + Cl^- = AgCl + e \qquad (1.29)$$

takes place at the electrode. Thus, instead of a positive charge passing into the solution, a negative charge passes from the solution into the electrode. Nevertheless, the result is the same: the amount of chloride in vessel 1 decreases by Q/F (see Fig. 27). The fact that the decrease in the KCl content in vessel 1 is only $0.49 \, Q/F$ can be explained by the transport of $0.51 \, Q/F$ mol Cl^- from vessel 2 to vessel 1. At the same time reaction (1.29) occurs at the electrode of vessel 2 in the opposite direction. However, the increase in the amount of KCl in vessel 2 is only $0.49Q/F$ because just this amount of potassium ions has passed from vessel 1 to vessel 2 and $0.51Q/F$ of chloride ions has been transported in the opposite direction. If our observation is restricted to the processes in the capillary, the total charge Q is transported partly in the direction from vessel 1 to vessel 2 by cations as fraction $0.49Q$, and partly in the opposite direction by anions as $0.51Q$. If the charge were also carried, for example, by electrons (fraction xQ) the amount of charge transported by the potassium ions would decrease to $0.49 \, (1 - x)Q$ and that transported by chloride ions to $0.51(1 - x)Q$ which, of course, is not in accord with observation.

Thus, since the electrical charge is transported only by potassium and chloride ions, the electrolyte conductivity is directly proportional to the concentration of KCl (if this is not excessively high). In contrast to a metallic conductor, the conductivity of an electrolytic conductor increases with temperature, because at higher temperatures the ions move more rapidly. The mobility of various types of ions is different. As we have already seen, in a potassium chloride solution the potassium ions transport 49% of the overall charge transferred. The ratio between the charge transported by an individual type of ion to the total charge transferred is termed the transport number t_i. Thus, $t_{K^+} = 0.49$ and $t_{Cl^-} = 0.51$. Potassium chloride is an exceptional case where the transport numbers of the

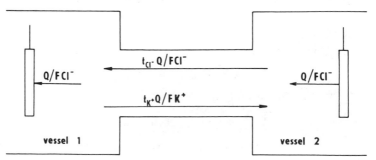

Fig. 27. Charge transport by cations and anions in a tube

cation and of the anion are almost equal. On the contrary, e.g. for lithium chloride, $t_{Li^+} = 0.33$ and $t_{Cl^-} = 0.67$.

The transport numbers reflect the relative speed of the ions. We will now consider the absolute values of ion mobilities. The electric current flowing through the capillary is related to the current density j by the equation

$$I = Aj$$

where A is the cross-section of the capillary. The current density is composed of the flux of cation J_B and of anion J_A

$$j = F(J_B - J_A)$$

The minus sign indicates that the anions move in an electrical field in the opposite direction to the cations. This equation is a formulation of the *Faraday Law* which governs electrolytic charge transport or *migration* (i.e. the transport of charge linked to the transport of matter). In a more general formulation, when the charge number of the cation is z_B and that of the anion z_A

$$j = F(z_B J_B + z_A J_A) \tag{1.30}$$

The flux of an ion is directly proportional to the ion concentration at a given point, to the charge number, and to the 'driving force' of the transport which is, in the present case, the electrical field strength. The resulting product must be multiplied by the Faraday constant to convert quantities related to the transport of electricity to the transport of matter. Here the electrical field strength X simply is equal to the ratio of the voltage or, more properly, the electrical potential difference, acting between the ends of the capillary and its length $\Delta V / l$, multiplied by -1. The fluxes of the cation J_B and of the anion J_A are then given by the equation

$$\begin{aligned} J_B &= U_B F c_B z_B X = -U_B F c_B z_B U/l \\ J_A &= U_A F c_A z_A X = -U_A F c_A z_A U/l \end{aligned} \tag{1.31}$$

The proportionality constants U_B and U_A are the *mobilities* of ions B and A. The mobilities reflect the ease with which a particle overcomes the resistance of the medium to its movement. A spherical particle with a radius r moving in a fluid medium with a viscosity η under the influence of a force f will be considered. The velocity v of that particle is given by the Stokes Law (cf. p. 16)

$$v = f/(6\pi\eta r)$$

The flux of ion B is related to the amount of the substance transported, so that

$$J_B = N_A v = N_A f/(6\pi\eta r) = N_A (6\pi\eta r)^{-1} c_B z_B F X$$

where N_A is the Avogadro constant. Thus

$$U_B = N_A/(6\pi\eta r) \tag{1.32}$$

This simple relationship is based, of course, on a gross simplification. The principal assumption of the Stokes Law is that the moving particle is much larger than the molecules of the medium. This is certainly not true for small ions moving, for example, in water. Nevertheless, more complete theories yield equations containing only numerical factors other than 6π present in the simple Stokes Law.

By combining equations (1.30) and (1.31) the current density equation

$$j = (U_B c_B z_B^2 + U_A c_A z_A^2)F^2 X \tag{1.33}$$

is obtained. Ohm's Law can be written in the form

$$I = -G\Delta V$$

where G is the conductance. (This equation contains a minus sign because the positive current flows in the opposite direction to the electrical potential difference.) Dividing and multiplying by the length of the conductor l and dividing by its cross-section A yields

$$I/A = -(Gl/A)\cdot(\Delta V/l)$$

or

$$j = \kappa X \tag{1.34}$$

where κ is the *conductivity* of the conductor.

Combining equations (1.33) and (1.34) yields the conductivity of an electrolyte

$$\kappa = (U_B c_B z_B^2 + U_A c_A z_A^2)F^2 \tag{1.35}$$

The formula of an electrolyte consisting of cations with charge number z_B and anions with charge number z_A is $B_m A_n$ where, considering the electroneutrality of the electrolyte

$$mz_B + nz_A = 0$$

Because, for the overall concentration c of $B_m A_n$, it holds that

$$c = c_B/m = c_A/n$$

Equation (1.35) can be then written in the form

$$\begin{aligned}\kappa &= (U_B z_B - U_A z_A)mz_B F^2 c \\ &= (U_B z_B - U_A z_A)nz_A F^2 c\end{aligned} \tag{1.36}$$

If the conductivity is divided by concentration c the *molar conductivity*

$$\Lambda = \kappa/c \tag{1.37}$$

is obtained.

Thus, the unit of molar conductivity is, for example, $S.mol^{-1}.m^2$. However, for the conductivity κ the unit is $S.cm^{-1}$ and for the concentration the usual unit $mol.dm^{-3}$. Then equation (1.37) must be written with an inconvenient conversion factor, $\Lambda = 1000\kappa/c$, to yield the simple unit $S.mol^{-1}.cm^2$. This discussion shows the advantage of expressing all quantities in an equation in coherent units. The simple, but not always the most practical, procedure would be to use only basic SI units (see Appendix D).

The molar conductivity can be separated into *ion conductivities* corresponding to the individual ions that form the electrolyte BA

$$\Lambda = m\lambda_B + n\lambda_A \tag{1.38}$$

Using tabulated ion conductivities (see Table 7), the molar conductivity (at least in a rather dilute solution) of a given electrolyte can be calculated from equation (1.37).

Similarly to the equilibrium properties, the transport properties are also influenced by electrostatic and other effects which appear at millimolar and higher concentrations. The main electrostatic effect on the conductivity is a result of the interaction between the ion atmosphere and the ion enclosed by the ion atmosphere. The ion atmosphere has larger dimensions than the ion and moves in the opposite direction. Its movement then decelerates the moving ion and its contribution to the overall conductivity decreases. This phenomenon is called the *electrophoretic effect*. In addition, it seems as if the ion would escape from the centre of the ion atmosphere which would have to reform continuously around it. This reformation of the ion atmosphere requires a fraction of time, termed the *relaxation time*. During this time period the ion lies outside the centre of the ion atmosphere, which results in a force directed against the movement of the ion. The rather complicated theory developed by Onsager explained the empirical law found much earlier by Kohlrausch

$$\Lambda = \Lambda^0 - k\sqrt{c} \tag{1.39}$$

where Λ^0 is the molar conductivity at infinite dilution, k an empirical constant, and c the concentration of the electrolyte.

The experimental arrangement shown in Fig. 25 is only rarely used for determination of conductivities. Conductometric cells of the type shown in Fig. 28 contain platinum electrodes covered with a layer of porous platinum ('platinized platinum') to increase the surface of the electrodes. Instead of a d.c. current an a.c. current signal is fed to the electrodes and the measurement is carried out using a conductivity bridge (see Fig. 29).

Table 7   Ion conductivities

Ion	Ion conductivity ($S\,cm^2\,mol^{-1}$)	Ion	Ion conductivity ($S\,cm^2\,mol^{-1}$)
H_3O^+	349.8	OH^-	198.3
Li^+	38.7	F^-	55.4
Na^+	50.1	Cl^-	76.4
K^+	73.5	Br^-	78.1
Rb^+	77.8	I^-	76.8
Cs^+	77.3	SO_4^{2-}	160.0
Ca^{2+}	119.0	$Fe(CN)_6^{3-}$	302.7

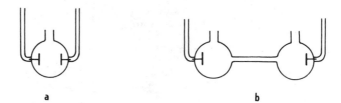

Fig. 28. Conductometric cells for (a) low and (b) high conductivity solutions

The rate of transport processes (they include diffusion and ion migration, and also other processes such as heat conduction, i.e. transport of energy) is directly proportional to the driving force of a given process and, in the case of the transport of matter, to the number of particles under the impact of that force. The driving force comes from the deviation of the system from equilibrium. The system has a larger tendency to return to equilibrium at greater deviations. An attempt will now be made to describe this situation more quantitatively.

The energy of 1 mol of any uncharged component of the solution is characterized by the chemical potential, i.e. the magnitude of the Gibbs energy corresponding to 1 mol of this component (this approach is, however, not restricted to solutions). The Gibbs energy is a measure of the ability of the system to carry out a definite amount of 'useful work' (see Appendix B). In the present case the useful work can be osmotic work (cf. p. 22). When the solution is separated, by a membrane permeable for a given component, from another

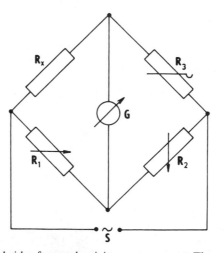

Fig. 29. A Wheatstone bridge for conductivity measurement. The measured resistance R_x is inserted in the bridge circuit with variable resistances R_1, R_2, and R_3. If the bridge is balanced (electronic balance detector G indicates zero) then according to Kirchhoff's laws the unknown resistance is given by $R_x = R_1 R_3 / R_2$. S is the electronic source of sinusoidal voltage. The cell is standardized using a solution of a known conductivity, κ_s, so that a standard conductance G_s is determined. The conductivity of an investigated solution is then given by $\kappa = \kappa_s \cdot G/G_s$, where $G = 1/R_x$

solution containing a lower concentration of that component, the difference in the osmotic pressures can be used for mechanical work (for example, to lift a piston placed in the vertical part of the tube of Fig. 14).

When the chemical potential differs between two positions in the solution, this solution is not in equilibrium. The driving force of diffusion tends toward returning to equilibrium. In a dilute solution the chemical potential of a given component i is given by the equation

$$\mu_i = \mu_i^0 + RT \ln c_i \tag{1.40}$$

where μ_i^0 is the standard chemical potential independent of the concentration of component i in the solution and c_i is the concentration of this component.

When the ions are present in the solution, their electrical energy must be used properly to characterize their energy. The electrical energy of an ion is given by the product of the charge of 1 mol of the ions $z_i F$ and the electrical potential at the position of the ion in the solution

$$\mu_{e,i} = z_i F \varphi$$

The *electrical potential* (or, more exactly, the *inner electrical potential*) φ is given by the work connected with bringing a unit charge (1 C) from infinity to the given position (here, to the given point in the solution). (This quantity cannot be measured directly.)

The sum of the chemical potential and of the electrical contribution $\mu_{e,i}$ is termed the *electrochemical potential*

$$\tilde{\mu}_i = \mu_i + \mu_{e,i} = \mu_i^0 + RT \ln c_i + z_i F \varphi \tag{1.41}$$

In an electrolyte solution the driving force originates from different values of the electrochemical potential between different positions in the system. When the electrochemical potential only varies along a single coordinate, x, the driving force for transport can be usefully identified with the differential quotient of the electrochemical potential with respect to this coordinate (i.e. with the gradient of the electrochemical potential)

$$X_i = d\tilde{\mu}_i/dx$$

The transport process characterized by the gradient of the electrochemical potential is simultaneously occurring diffusion and ion migration (sometimes termed electrodiffusion for both processes together). The rate of this transport process can be expressed simply as the flux of the substance proportional to the number of particles (or concentration c_i) and to the driving force with the mobility of the substance U_i as the proportionality constant

$$
\begin{aligned}
J_i &= -U_i c_i \, d\tilde{\mu}_i/dx \\
&= -U_i RT \, dc_i/dx - U_i z_i c_i F \, d\varphi/dx
\end{aligned} \tag{1.42}
$$

This relationship is called the Nernst–Planck equation. Obviously it is a combination of the Fick Law (1.26) and the equation of ion migration (1.31). The diffusion coefficient D_i (cf. equation (1.26)) is related to the mobility U_i by the equation

$$D_i = RTU_i \qquad (1.43)$$

and the electrical field strength X is equal to the gradient of the electrical potential multiplied by -1

$$X = -d\varphi/dx$$

Sometimes it is useful to introduce the term *electrolytic mobility*

$$u_{i_\bullet} = |z_i| FU_i$$

Obviously

$$D_i/u_i = RT/(|z_i|F) \qquad (1.44)$$

which is the Nernst–Einstein equation, valid, however, only for dilute solutions.

The Nernst–Planck equation (1.42) relates the diffusion and the ion-migration by a single constant, U_i, characterizing the individual properties of a given type of ion. At this stage we shall be able to compare the rates of the two processes. Let us return to the experiment on steady state diffusion (Fig. 22) and ion migration (Fig. 25). In the diffusion experiment 10^{-3} M KCl is in compartment 1 and 2 $\times 10^{-3}$ M KCl in compartment 2. In the migration experiment the same concentration, 10^{-3} M KCl, is in both compartments. The length of the capillary is 0.5 cm and the temperature of the solutions 25°C (298 K). For the sake of simplicity it is assumed that both the diffusion coefficient of K^+ and of Cl^- have the same value, i.e. 2×10^{-5} cm^2.s^{-1}. In order to obtain the result in coherent units we must recalculate the concentrations to units of mol.cm^3 (1 M KCl $= 10^{-3}$ mol.cm^{-3} KCl). From equations (1.26) and (1.27) we have for the diffusion flux $J_{KCl} = -4 \times 10^{-11}$ mol.cm^{-2}.s^{-1}. Now determine the magnitude of the voltage which must be imposed on the silver–silver chloride electrodes to obtain the same migration flux of potassium ions (the flux of chloride ions has the same absolute value but the opposite sign). Equations (1.31) an (1.43) yield the relationship

$$U = lJ_K + (U_K + c_K + F)^{-1} = RTlJ_{K^+}(D_{K^+}c_{K^+}F)^{-1}$$
$$= 0.026 \text{ V}$$

Thus, in the absence of extreme concentration gradients in the solution, transport by ion migration at electrical field strengths of an order of volts per centimeter is much faster than diffusion.

In the next experiment the experimental arrangement shown in Fig. 22 will again be used, but compartments 1 and 2 will be filled with an electrolyte solution whose cation and anion have very different transport numbers. Thus, vessel 1 contains 10^{-4} M HCl and vessel 2, 10^{-3} M HCl. Once diffusion has attained a steady state, the electrical potential difference between the vessels is measured (for example, by two silver–silver chloride electrodes connected to each of the solutions by a capillary tube filled with a saturated KCl solution, see p. 76). The voltage between the electrode placed in vessel 2 and the electrode in vessel 1 measured by an electronic voltmeter is approximately $-38\,mV$ (the electrode placed in vessel 2 being negative with respect to the electrode placed in vessel 1).

Thus, electrolyte diffusion alone can produce electrical potential differences in the solution. Obviously, ions of higher mobility (here H_3O^+) tend to outrun the ions with lower mobility (Cl^-). This cannot result in separate accumulation of cations or anions because electroneutrality must be preserved. However, the fast and the slow ions interact electrostatistically to produce an electrical potential difference in the solution. This *diffusion potential* is the quantity we have measured. Diffusion potential is a general term for electrical potential changes accompanying electrolyte diffusion. In our experiment (and very often in potentiometric measurements, see p. 81) the vessels where the diffusion is negligible are connected with a junction where it actually takes place. The potential difference between the solutions is then called the *liquid junction potential* $\Delta\varphi_L$.

Combining equations (1.42) and (1.30) yields the result for $j = 0$ (no electrical current flows through the solution from an external source)

$$\frac{d\varphi}{dx} = -\frac{RT}{F}\frac{U_{H_3O^+} - U_{Cl^-}}{U_{H_3O^+} + U_{Cl^-}}c^{-1}\frac{dc}{dx}$$

The equation is easily integrated, bearing in mind that the electrical potential in vessel 2 (HCl concentration c_2) is φ_2 and in vessel 1 (concentration c_1) φ_1

$$\varphi_2 - \varphi_1 \equiv \Delta\varphi_L$$
$$= -\frac{RT}{F}\left[\frac{U_{H_3O^+} - U_{Cl^-}}{U_{H_3O^+} + U_{Cl^-}}\right]\ln\frac{c_2}{c_1}$$

Since the transport numbers for a symmetrical electrolyte (with $z_B = -z_A$) are defined as

$$t_B = \frac{U_B}{U_B + U_A}; \qquad t_A = \frac{U_A}{U_B + U_A} \tag{1.45}$$

then

$$\Delta\varphi_L = \frac{RT}{F}(t_{H_3O^+} - t_{Cl^-})\ln\frac{c_2}{c_1} \tag{1.46}$$

This equation will be specified for $t = 25°C$ using decadic logarithms. For the coefficient $2.303\ RT/F$ ($R = 8.314$ J.K^{-1}.mol^{-1}, $T = 298\,K$, $F = 96485$ C.mol^{-1}), 0.0591 V is obtained. (It is useful to remember this value since it will frequently be encountered in Chapters 2 and 3.) Thus

$$\Delta\varphi_L = -0.0591\,(t_{H_3O^+} - t_{Cl^-})\ln\frac{c_2}{c_1}$$

The transport number of the hydronium ion in the HCl solution is $t_{H_3O^+} = 0.82$. Thus $\Delta\varphi_L = 38\,mV$, in accordance with the experiment.

Equation (1.44) can be deduced without making assumptions on the distribution of diffusing ions in the capillary. On the other hand, when the solutions in contact contain several electrolytes, this simple procedure cannot be used and some information on ion distribution is necessary for the determination of the liquid junction potential. When, for example, the solutions in both the vessels brought into contact are connected by turning a stopcock (see Fig. 30) the two solutions are mixed together in the junction so that the concentrations at one end of the junction can be expected to linearly increase or decrease toward the other end of the junction. This model of a liquid junction was treated mathematically by Henderson. Both the solutions in contact contain ions 1, 2, 3, etc. with charge numbers z_1, z_2, z_3, etc. and mobilities U_1, U_2, U_3, etc. Their concentrations in vessel 1 are $c_1(1), c_2(1), c_3(1)$, etc. and in vessel 2, $c_1(2), c_2(2),$ $c_3(2)$, etc. The Henderson formula gives the relationships for the liquid junction potential

$$\Delta \varphi_{\text{L}} = -\frac{RT}{F}$$

$$\frac{z_1 U_1 [c_1(2) - c_1(1)] + z_2 U_2 [c_2(2) - c_2(1)] + z_3 U_3 [c_3(2) - c_3(1)] + \ldots}{z_1^2 U_1 [c_1(2) - c_1(1)] + z_2^2 U_2 [c_2(2) - c_2(1)] + z_3^2 U_3 [c_3(2) - c_3(1)] + \ldots}$$

$$\times \ln \frac{z_1^2 U_1 c_1(2) + z_2^2 U_2 c_2(2) + z_3^2 U_3 c_3(2) + \ldots}{z_1^2 U_1 c_1(1) + z_2^2 U_2 c_2(1) + z_3^2 U_3 c_3(1) + \ldots} \qquad (1.47)$$

In our simple experiment the Henderson formula simplifies to equation (1.44).[2]

In the next experiment the solution in vessels 1 and 2 will contain 0.1 M KCl as well as 10^{-4} M and 10^{-3} M HCl. The voltmeter connected to the silver–silver chloride electrodes indicates zero. Excess potassium chloride acting as an indifferent electrolyte (cf. p. 29) prevents the formation of a diffusion potential. At any position in the electrolyte the ions of the indifferent electrolyte are present in a much larger concentration than the diffusing ions. They suppress the electrostatic interaction between the diffusing ions so that no appreciable electrical field is

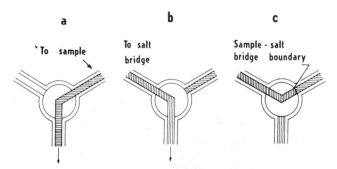

Fig. 30. Liquid junction with free diffusion: a, the test solution is drawn (in the direction of the arrow) into the stopcock; b, the saturated KCl solution from the liquid bridge is drawn into the stopcock; c, by turning the stopcock the test solution/liquid bridge liquid junction is formed. According to G. Mattock and D. M. Band

formed. The same conclusion follows from equation (1.47) when we identify ions 1 and 2 with the cation and the anion of the indifferent electrolyte. Since the concentration of the other ions can be neglected in comparison with c_1 and c_2, the fraction in the logarithm becomes equal to unity.

So far a macroscopic picture of the transport processes has been considered. Continuous distribution of the content of the substances in the solvent medium has been assumed and the molecular point of view, i.e. the behaviour of individual ions in the solution, has been ignored. The former approach is termed phenomenological.

When the motion of individual ions is considered it cannot be assumed that they move straight in the direction of the drop in concentration or in the direction (eventually against the direction according to the sign of their charge) of the electrical field. In fact, they are located in the cavities between the solvent molecules, where they vibrate. After several hundred vibrations they acquire sufficient energy in a random event to jump into a neighbouring stable position where the same process occurs again. In this way the ions move irregularly, as shown in Fig. 31. As a result of this random motion the particle covers an average distance of about several hundredths of a millimeter in a second. Einstein and Smoluchowski worked out the statistical theory of diffusion and, in connection with this theory, deduced the expression for the average distance $\bar{\Delta}$ that a particle covers during observation time τ

$$\bar{\Delta} = \sqrt{2D\tau} \tag{1.48}$$

where D is the diffusion coefficient of the particle (see Fig. 31).

When the solution contains a constant concentration of the solute this translocation of particles is of no importance for any total change. (In this paragraph the diffusing species are termed particles because the discussion is the same for ions as for uncharged molecules.) When the particle leaves position 1 another particle subsequently moves to this position. However, new circumstances prevail when a concentration gradient is formed in the solution.

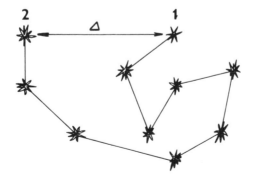

Fig. 31. Irregular particle motion in solution. The particles oscillate about a mean position for some time and then jump to a neighbouring position. Δ denotes the displacement of the diffusing particle during time τ

The probability of a particle moving in the direction of increasing concentration is the same as in the direction of decreasing concentration. However, the probability for particles to be transported from the region of higher concentrations is greater than for those coming from the region of lower concentrations. This results in the prevalent transfer of the solute in the direction of the concentration drop, as described by the Fick Law.

When an external source of electricity produces an electrical field in the solution, ion migration begins. Movement in the direction of the field or against it results from the slight preference among the cations for movement in the direction of the field (among the anions in the opposite direction). An electrical field of 1 V.cm^{-1} will be considered which can easily be realized in the experiment shown in Fig. 25. This field results in a component of the velocity of a cation in the direction of the field of about $5 \, \mu m.s^{-1}$. The velocities of the ions when they move from one stable position to another are approximately 10^7 times larger, so that when an ion makes about 10^{11} jumps per second the external electrical field produces a shift corresponding to only 10^4 jumps per second. This is a minute influence compared with the thermal motion of the ions (this situation is analogous to the conduction of electricity in a metal, see p. 7). Nevertheless, the influence of the field on the oriented motion of the ions is quite marked.[3]

The ion conductivity (see Table 7) and, therefore, also the diffusion coefficient of the hydronium ion, is much larger than those of other ions, although the hydronium ion is not that much smaller. This striking phenomenon results from the fact that, in the translocation of the hydronium ion, the whole structure is not transferred, only an arbitrary proton from the hydronium ion. Consequently

$$\begin{array}{c} H \\ \diagdown \\ \diagup \quad O-H^+ \longrightarrow O \\ H \end{array} \begin{array}{c} \diagup H \\ \diagdown \\ H \end{array}$$

Proton transfer from one water molecule to another requires a certain activation energy (see Appendix C). With light particles such as the electron, similar transfer processes occur by the tunnel mechanism: the electron wave function extends across the barrier which inhibits the electron transfer and, therefore, a certain probability exists that the transfer occurs without supplying activation energy to the electron (see also p. 98). As shown by Bernal and Fowler, under certain conditions the proton can also be transferred by a tunnel mechanism (of course, to shorter distances than the electron). If, however, proton transfer in water occurred by a pure tunnel mechanism, the ion mobility and the diffusion coefficient would be much larger than the values found experimentally. Obviously the transfer rate is not in fact determined by the actual velocity of proton motion, which is very large, but by the rotation of the water molecule to a position in which it can accept the proton[4](see Fig. 32).

References

1. See, for example, J. Koryta, J. Dvořák, and V. Boháčková, *Electrochemistry*, Chapman and Hall, London, 1973.
   V. G. Levich, *Physico-Chemical Hydrodynamics*, Prentice-Hall, Englewood Cliffs, 1962.

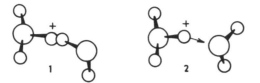

Fig. 32. Proton transfer from one water molecule to another requires suitable orientation of both molecules. 1, transfer is not possible; 2, transfer is possible

2. G. Eisenman, *Glass Electrodes for Hydrogen and Other Cations: Principles and Practice*, M. Dekker, New York, 1967.
   J. Koryta, *Ion-selective Electrodes*, Cambridge University Press, Cambridge, 1975.
3. C. A. Vincent, *J. Chem. Educ.*, **53** (1976) 490.
4. B. E. Conway, 'Proton solvation and proton transfer processes in solution', in: *Modern Aspects of Electrochemistry*, Vol. 3 (J. O'M. Bockris and B. E. Conway, Eds.), Butterworths, London, 1964.

Chapter 2
Electrodes

Oxidation and Reduction at Electrodes

Electrolysis has been already mentioned on p. 10, here an experiment with electrolysis in a modified arrangement will be described. The *electrolytic cell* (Fig. 33) will consist of a glass vessel containing a large-surface platinum electrode (for example, a plate with an area of 1 cm²) and a small platinum electrode (for example, a platinum wire with a diameter of 0.1 mm sealed in a glass tube so that a length of 0.5 mm projects). The surface area of the small electrode (microelectrode) is 1.6×10^{-3} cm². The electrolyte solution in the cell is 10^{-3} M $Fe(ClO_4)_2$, 10^{-3} M $Fe(ClO_4)_3$, and 0.1 M $HClO_4$. Both electrodes are connected to a suitable voltage source (cf. Fig. 25). In the present case an electronically controlled low-output impedance voltage source is used instead of a simple potentiometer. The current-measuring device, which is a galvanometer in Fig. 25, will record even rather rapidly changing current. Thus, a fast pen

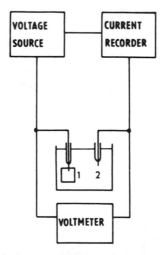

Fig. 33. A simple electrolysis arrangement. A large-surface electrode 1 and a microelectrode 2 are polarized by a voltage source. The current is measured with a recording ammeter or oscilloscope as a function of time and the electric potential difference between the electrodes is measured with a recording voltmeter

recorder or an oscilloscope must be used. These instruments record the electric current (vertical axis) as a function of time (horizontal axis).

When zero voltage ($\Delta V = 0$) is imposed on the electrodes, no current deflection is indicated by the recorder. Obviously, the system is in equilibrium. Assume that a small voltage (for example, 5 mV) is applied to the electrodes, *polarizing* them so that the microelectrode is positive. The recorder first indicates a very high and rapidly decaying positive current. Later the current drops rather slowly (Fig. 34). When the voltage source is switched off and the electrical potential difference between the electrodes is measured by a recording electronic voltmeter (no current flows through the instrument because of its high input resistance), the initial value of 4 mV falls slowly to zero (Fig. 35).

The time-course of the current during switching-on of the external voltage source (i.e. during electrolysis) as well as the time-course of the electrical potential difference between the electrodes after switching-off the voltage source, are caused by several processes which take place at the electrodes, in their surroundings, and in the bulk electrolyte. Because of the much smaller size of the microelectrode compared with the large electrode, the current density at the microelectrode is many times (three orders of magnitude) smaller than at the other electrode. Therefore the microelectrode, where the effect of imposed voltage or *polarization* is apparent, is often called the polarized (or polarizable) electrode, while the large electrode where this effect is negligible is *non-polarized*.

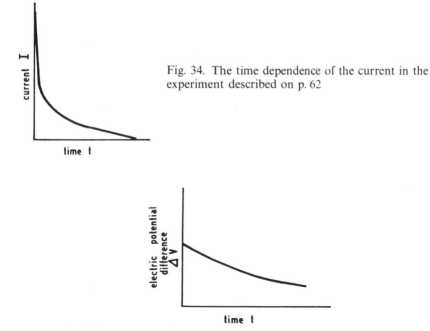

Fig. 34. The time dependence of the current in the experiment described on p. 62

Fig. 35. The time dependence of the electric potential difference between the electrodes after switching off the external electricity source in the experiment described on p. 62

As will be discussed in more detail below, the electrode has two basic properties. It is an acceptor or donor of electrons (or holes, eventually) which are transferred from the solution or into the solution. On the other hand, when it is in contact with an electrolyte solution, it bears a definite charge uniformly spread over its surface. The charge of the electrode atracts ions of opposite sign from the solution and, in this way, a molecular condenser—the electrical double-layer—is formed. Its capacity is enormous, since the distance between the surface of the electrode and the plane passing through the centres of the ions at closest approach to the electrode surface is about 0.1 nm (cf. equation (1.2)). When the electrical potential difference between the electrodes is changed, the voltage at this condenser is also changed and a certain charge must be supplied to (or taken from) the electrode. The abrupt increase of current followed by a rapid drop (see Fig. 34) corresponds to the charging of the electrode. This kind of current is termed the *condenser* or *charging current*. At a later stage the current is connected with the first of the main properties of the electrode since it is caused by the electrode reaction

$$Fe^{2+} \underset{k_{red}}{\overset{k_{ox}}{\rightleftarrows}} Fe^{3+} + e \tag{2.1}$$

and consequently by a process of electron transfer between the electrode and the solution. The current originating from this process reaches a maximum value at the instant when the voltage from the external source is switched on. This maximum value is given by the rate of reaction (2.1) which, when referred to the unit area of the electrode, has the value (usually in units $mol.cm^{-2}.s^{-1}$)

$$v = k_{ox}c_{Fe^{2+}} - k_{red}c_{Fe^{3+}} \tag{2.2}$$

In this equation k_{ox} is the rate constant of the oxidation reaction (i.e. of reaction (2.1) occurring in the direction from left to right), k_{red} is the rate constant of the reduction reaction (when reaction (2.1) proceeds in the opposite direction), and $c_{Fe^{2+}}$ and $c_{Fe^{3+}}$ are the concentrations of divalent and trivalent iron respectively. As we shall see below, the constants k_{ox} and k_{red} depend on the electrical potential (the proper definition of this electrical potential will also be considered below). In the course of the oxidation reaction, the ions of ferrous iron are depleted in the surroundings of the electrode while the amount of ferric iron increases and their concentrations at the electrode surface change compared with their values at a larger distance from the electrodes. The concentration gradients formed in this way result in diffusion transport of ferrous ions to the electrode and of ferric ions from the electrode. These ions could be also transported in the same direction by migration (in our experimental arrangement, however, the contribution of migration to the total transport process is small because of the presence of the excess indifferent electrolyte, $HClO_4$). The transport cannot, in fact, balance the concentration changes at the electrode completely, therefore reaction (2.1) slows down.

When the external voltage is switched off, first an instantaneous decrease by 1 mV in the potential difference between the electrodes is observed. This sudden decrease in the voltage is due to the disappearance of the *ohmic drop in the potential* in the solution. The electrolytic cell acts as an electrical resistance, R, and when an electric current I flows through it, a part equal to IR of the voltage imposed on the electrodes is lost, mainly owing to the resistance of the electrolyte. The subsequent slow drop of the electrical potential difference is caused by the diffusion of ferric ions from the electrode and of ferrous ions to the electrode. Obviously, the potential difference in a currentless state depends on the concentrations of both ferric and ferrous ions, in general on the concentrations of the oxidant and of the reductant form of the species reacting at the electrode.

The reason for the electron transfer from the ferrous ion to the electrode can be found by consideration of the band structure of the electrode and similar characteristics of the oxidant and of the reductant in the solution. The average value of the energy of an electron in the solid phase is fixed by the energy of the Fermi level ε_F (cf. p. 6). This energy can also be expressed by the electrochemical potential of the electron in the metallic phase $\tilde{\mu}_e$. The energy of the Fermi level is usually referred to a single electron, whereas the electrochemical potential corresponds to a unit amount (one mol) of electrons. An attempt will now be made to find a relationship for $\tilde{\mu}_e$ in the solution. The oxidant Fe^{3+} and the reductant Fe^{2+} which are present do not peacefully coexist, but unceasingly react together. Electrons are transferred from the particles of the reductant to those of the oxidant so that the original Fe^{3+} becomes Fe^{2+} and vice versa. Chemically this *exchange process* is expressed by the equation

$$Fe^{2+} \rightarrow Fe^{3+} + e$$
$$Fe^{3+} + e \rightarrow Fe^{2+} \tag{2.3}$$

Thus, in the solution only a more or less fictitious concentration of electrons is present that corresponds to the electrons individually existing during the jumps from Fe^{2+} to Fe^{3+}. The situation is analogous to the existence of free protons in water. This minute concentration of electrons in the solution is determined by the highest energy level of the electron in the Fe^{2+} ion (also called the highest donor level or term), by the lowest non-occupied energy level of the electron in the Fe^{3+} ion (by the lowest acceptor term) and by the number of these levels which are available in a unit volume of the solution, i.e. by the concentration (more exactly, by the activity) of the Fe^{2+} and Fe^{3+} ions. Consequently, the Fermi level of the electron is situated between the highest occupied energy level of the electron in the reductant form (Fe^{2+}) and the lowest non-occupied energy level in the oxidant form (Fe^{3+}) of an oxidation–reduction system (see Fig. 36). This can be quantitatively described by the energy balance of the electrochemical potentials of the participating species in the solution.

For chemical reactions a general equation

$$aA + bB + \cdots = cC + dD + \cdots \tag{2.4}$$

can be written. The upper case letters denote the molecules and other particles (the electrons, for example) participating in a reaction, and the lower case letters the stoichiometric coefficients. In reaction (2.3) all the stoichiometric coefficients

66

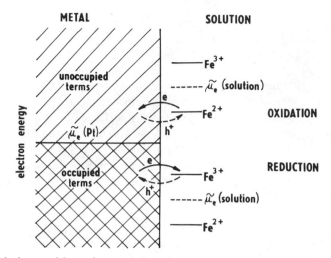

Fig. 36. Relative position of occupied and unoccupied electron terms in a metallic electrode and in the oxidized and the reduced forms in solution for oxidation and reduction. The relative position of the Fermi levels (expressed as electrochemical potentials of electrons $\tilde{\mu}_e$) decides whether an electron is transferred from the solution into the electrode (or a hole from the electrode into the solution) or in the opposite direction. According to H. Gerischer

are equal to unity. The portion of the Gibbs energy attributed to a unit amount of reacting particles is the chemical (in the case of charged particles, the electrochemical) potential μ_i. At equilibrium (cf. Appendix B) these chemical potentials are related by an equation analogous to (2.4)

$$a\mu_A + b\mu_B + \cdots = c\mu_C + d\mu_D + \cdots$$

For the first of equations (2.3)

$$\tilde{\mu}_{Fe^{3+}} = \tilde{\mu}_{Fe^{2+}} + \tilde{\mu}_e \qquad (2.5)$$

Considering equation (1.41), it holds that

$$\begin{aligned}
\tilde{\mu}_e(\text{solution}) &= \tilde{\mu}_{Fe^{2+}} - \tilde{\mu}_{Fe^{3+}} \\
&= \tilde{\mu}^0_{Fe^{2+}} + RT \ln a_{Fe^{2+}} + 2F\varphi(\text{solution}) \\
&\quad - \mu^0_{Fe^{3+}} - RT \ln a_{Fe^{3+}} - 3F\varphi(\text{solution}) \qquad (2.6) \\
&= \mu^0_{Fe^{2+}} - \mu^0_{Fe^{3+}} - RT \ln (a_{Fe^{3+}}/a_{Fe^{2+}}) - F\varphi(\text{solution})
\end{aligned}$$

Thus, a relationship is obtained among the electrochemical potential of the electron in solution, the activities of ferric and ferrous iron, and the inner electrical potential of the solution $\varphi(\text{solution})$. This last quantity is not suitable since it can be determined only approximately under some simplifying assumptions (see p. 76). In the present case, however, it can easily be eliminated.

The large platinum electrode which has been called 'non-polarizable' is virtually in equilibrium with the adjacent solution. An equilibrium of two phases in contact is characterized by the condition that the electrochemical potential of a particle in one phase is equal to the electrochemical potential of the same particle in the other phase (when the particle is, in fact, present in this phase). For the electrochemical potential of the electron in the large platinum electrode (denoted 1) and in the solution it holds that

$$\tilde{\mu}_e(\text{Pt}, 1) = \tilde{\mu}_e(\text{solution}) \tag{2.7}$$

The electrochemical potential of the electron in metal M is given by the equation

$$\tilde{\mu}_e(\text{M}) = \mu_e^0(\text{M}) - F\varphi(\text{M})$$

This equation contains no term indicating a concentration or an activity of an electron in the solid phase, since this is constant and is included in the term $\mu_e^0(\text{M})$. The change in the electron concentration could be accomplished by charging the metal, but the concentration of excess electrons (or holes) is negligibly small in comparison with their overall concentration.

In the same way as equation (2.6), equation (2.7) is rearranged to give

$$\varphi(\text{solution}) = (\mu_{\text{Fe}^{2+}}^0 - \mu_{\text{Fe}^{3+}}^0 - \mu_e^0(\text{Pt}))/F$$
$$- (RT/F)\ln(a_{\text{Fe}^3}/a_{\text{Fe}^{2+}}) + \varphi(\text{Pt}, 1) \tag{2.8}$$

Thus, the potential $\varphi(\text{solution})$ has been replaced by a function of the electrical potential of electrode 1 $\varphi(\text{Pt}, 1)$. Consequently, the conditions at the interface between electrode 1 and the solution are controlled by the activities of both components of the oxidation–reduction system which determine the Fermi level for the electron in the metal. The conditions at polarizable electrode 2 are, however, different because, with the help of an external voltage source, the position of the Fermi level in the metal can be shifted higher or lower compared with that of the electrons in the solution (Fig. 36). Consider the case when the energy of the electrons in the electrode is lower than in the solution, i.e.

$$\tilde{\mu}_e(\text{Pt}, 2) < \tilde{\mu}_e(\text{solution}) \tag{2.9}$$

Here $\tilde{\mu}_e(\text{Pt}, 2)$ is the electrochemical potential of the electrons in the small platinum electrode. The electrons from the highest occupied level in the ferrous ion can jump to the non-occupied acceptor level in the metal (corresponding to the same energy as the Fermi level in the solution). Writing out the expressions for the electrochemical potentials in equation (2.9) yields

$$\varphi(\text{solution}) < (\mu_{\text{Fe}^{2+}}^0 - \mu_{\text{Fe}^{3+}}^0 - \mu_e^0(\text{Pt}, 2))/F$$
$$- (RT/F)\ln(a_{\text{Fe}^{3+}}/a_{\text{Fe}^{2+}}) + \varphi(\text{Pt}, 2)$$

After substituting from this equation into equation (2.9) the expected result is obtained

$$\varphi(\text{Pt}, 2) - \varphi(\text{Pt}, 1) > 0 \tag{2.10}$$

As mentioned above, the inner potential of a phase is inaccessible to measurement. Thermodynamic means cannot even be used to measure a difference in the inner potentials of two different phases (called the *Galvani potential difference*). However, a common potentiometric method can be used to determine the difference in the electrical potentials of two chemically identical phases, for example, made of the same metal. The resulting value of ΔV is exactly equal to the difference in the electrical potentials of the two phases, in the present case $\Delta V > 0$ (i.e. a positive voltage has been imposed on the small electrode with respect to the large electrode). Under these circumstances the electrons jump to the electrode from the Fe^{2+} ion which is oxidized. The electrical current originated in this way is termed *anodic* (it is due to the transfer of electrons from the solution into the electrode or, possibly, by the transfer of holes into the solution). The electrode functions as an *anode* (see Appendix A).

When, on the other hand

$$\tilde{\mu}_e(\text{Pt, 2}) > \tilde{\mu}_e(\text{solution})$$

the electrons tend to escape from the electrode into a free acceptor level in the solution. Such a level is present in the ferric ion and is the lowest non-occupied level with an energy close to that of the highest occupied level in Fe^{2+}. Thus, $\Delta V < 0$ and the electrons from the electrode jump to the ferric ions which are then reduced. The resulting current is called *cathodic* because it passes through the *cathode* which is the electrode from which the electrons enter the solution.

Evidently at the other, large, non-polarizable electrode the processes take place in the opposite direction, i.e. when the small electrode is an anode the large one is a cathode and vice versa. If this were not the case the electroneutrality of the whole system would be disturbed. However, the flow of current has a negligible impact on the non-polarizable electrode because the current density there is very low.

When, of course, $\Delta V = 0$, the whole system is in equilibrium and no current passes through it.

The deduction of long equations such as, for example, (2.8) would seem tedious when the resulting relationship (2.10) is meagre and almost self-evident. Fortunately, these relationships have a wider validity than for our simple case.

The strengths of various oxidants and reductants are different. Thus, for instance, the ion of trivalent iron, Fe^{3+}, oxidizes the ion of divalent vanadium, V^{2+}, according to the equation

$$Fe^{3+} + V^{2+} = Fe^{2+} + V^{3+} \qquad (2.11)$$

whereas it is unable to oxidize the ion of monovalent thallium Tl^+ to trivalent thallium Tl^{3+}. This is, however, easily oxidized by the ion of tetravalent cerium Ce^{4+}

$$2\,Ce^{4+} + Tl^+ = 2\,Ce^{3+} + Tl^{3+} \qquad (2.12)$$

Tetravalent cerium will also oxidize divalent vanadium. On the other hand, trivalent vanadium cannot oxidize the ferrous ion (equation (2.11) in the opposite direction). Trivalent thallium oxidizes divalent iron but will not react with trivalent cerium. Obviously, the ability of various oxidants and reductants is due to the energies of their lowest acceptor (non-occupied) levels or their highest donor (occupied) levels. Therefore, oxidations and reductions of various substances at electrodes will also require appropriate values of the Fermi levels. It would be suggested that the energy of the Fermi level at an electrode dipping in a solution containing the investigated oxidation–reduction system could be measured. Unfortunately, this is not that simple. A direct method for determination of the Fermi level of an isolated single electrode is not possible because, for such a measurement, two electrodes are necessary. When the experimental arrangement described on p. 62 is used then, at equilibrium, an electrical potential difference equal to zero would always be obtained because the electrodes are made of the same material and are placed in identical solutions. Let us try to change the material of electrode 1 by substituting platinum for gold. At equilibrium the energy of the Fermi level must be the same as in the preceding case, because it holds that

$$\tilde{\mu}_e(\mathrm{Au}) = \tilde{\mu}_e(\mathrm{solution}) = \tilde{\mu}_e(\mathrm{Pt}, 2) \qquad (2.13)$$

The interaction between electrons and metal ions in the metal lattice of gold is different from that in the metal lattice of platinum. Therefore

$$\mu_e^0(\mathrm{Pt}) \neq \mu_e^0(\mathrm{Au})$$

and

$$\varphi(\mathrm{Pt}, 1) \neq \varphi(\mathrm{Au}, 1)$$

However, the value of the electrical potential difference will again be equal to zero because the measuring device (a potentiometer or an electronic voltmeter) records the value of the voltage exclusively between two identical metallic phases, e.g. between two copper leads connecting the electrodes to the potentiometer. At equilibrium

$$\tilde{\mu}_e(\mathrm{Au}, 1) = \tilde{\mu}_e(\mathrm{Cu}, 1)$$
$$\tilde{\mu}_e(\mathrm{Pt}, 2) = \tilde{\mu}_e(\mathrm{Cu}, 2)$$

In view of equation (2.13) it can be concluded that the final potential difference is again zero. The substitution of one metal for another has resulted in no change, because the reaction taking place at the new electrode is again the same as in the preceding case, cf. reaction (2.1).

Obviously our system requires a radical change. For electrode 1 an electrode is chosen at which another reaction takes place than at electrode 2. The different reaction taking place at electrode 2 influences the corresponding Fermi level which will then be compared with that under the influence of the reaction occurring at electrode 1. This electrode is termed the *reference electrode*.

In aqueous solutions, which are of major concern in electrochemistry, hydrogen ions are always present. Therefore it is advantageous to use an electrode as a reference where a reaction occurs involving the participation of hydrogen ions. There are quite a few reactions of this type, but the simplest of them is

$$H_2 + 2H_2O = 2H_3O^+ + 2e \tag{2.14}$$

Here gaseous hydrogen is the reductant and the hydronium ion is the oxidant.

In order to obtain a suitable *hydrogen electrode*, platinum or another platinum group metal is used as the electrode material; for alkaline solutions porous nickel can also be used. For good function the platinum electrode must be covered by a thin layer of deposited platinum sponge called 'platinum black' (cf. p. 53). A hydrogen electrode is shown in Fig. 37.

At the hydrogen electrode gaseous hydrogen is oxidized to hydronium ions and, at the same time, hydronium ions are reduced to hydrogen, according to equation (2.14).*

The Fermi-level energy is again expressed in the form of the electrochemical potential of electrons in the solution,

$$\tilde{\mu}_e(\text{solution 1}) = \tfrac{1}{2}\mu_{H_2} + \mu^0_{H_2O} - \tilde{\mu}_{H_3O^+}$$
$$= \tfrac{1}{2}\mu^0_{H_2} + \mu^0_{H_2O} - \mu^0_{H_3O^+} + \tfrac{1}{2}RT\ln p_{H_2}$$
$$- RT\ln a_{H_3O^+} - F\varphi(\text{solution 1}) \tag{2.15}$$

In this equation the chemical potential of gaseous hydrogen is related to its relative pressure (or, to be exact, to its relative partial pressure in a gaseous mixture)

$$\mu_{H_2} = \mu^0_{H_2} + RT\ln p_{H_2}$$

Here, the relative pressure is given by the ratio of the actual pressure (unit Pa) to the standard pressure 1.01×10^5 Pa.

As with the majority of thermodynamic functions, an arbitrary constant term (which, of course, must be the same in a group of thermodynamic functions describing the same system) can be added to the standard chemical potential. Thus, it can be assumed that

$$\tfrac{1}{2}\mu^0_{H_2} + \mu^0_{H_2O} - \mu^0_{H_3O^+} = 0 \tag{2.16}$$

Considering equation (2.16) and relationship (1.22) ($\ln a_{H_3O^+} = 2.3\log a_{H_3O^+} = -2.3\,\text{pH}$) equation (2.15) takes the form

$$\tilde{\mu}_e(\text{solution 1}) = (RT/2F)\ln p_{H_2} + 2.3\,RTpH - F\varphi(\text{solution 1})$$

In order properly to carry out the measurement, the experimental arrangement must be modified. The gaseous hydrogen can reduce ferrous ions, but in an aqueous solution, in the absence of a catalyst, this reaction does not occur at all.

---

*The mechanism of this process is not simple and the final equilibrium between hydrogen, hydronium ions, and the electrode consists of a number of successive equilibrium steps. Gaseous hydrogen is in equilibrium with hydrogen dissolved in water. The dissolved hydrogen adsorbs on the platinum surface with formation of atomic hydrogen (for more details, see p. 100). The atomic hydrogen participates in the equilibrium with hydrogen ions and the final product, hydronium ions.

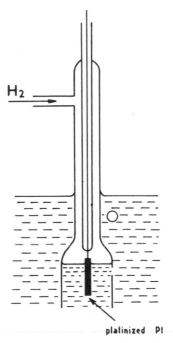

H_2

platinized Pt

Fig. 37. A hydrogen electrode

On the other hand, it can occur 'through the electrode'. The electrode will, at the same time, oxidize hydrogen and reduce Fe^{3+} (similar processes will be discussed in more detail in the section on mixed potentials on p. 108). Therefore the electrode, where hydrogen reacts, must be prevented from coming into contact with the ferric ions. This can be achieved using a permeable diaphragm separating the solution at electrode 2 (containing 0.001 M $Fe(ClO_4)_2$, 0.001 M $Fe(ClO)_3$, and 0.1 M $HClO_4$) from solution 1 which contains, for example, only 0.1 M $HClO_4$.

This cell is illustrated by a graphical scheme where the interface between the solid phases or between a solid and a liquid or gaseous phase is indicated by a vertical full line and the interface between the solutions by a vertical dashed line:

$$\underset{\text{Pt}}{\overset{1}{\vert}}\ \Big\vert\ H_2\ \Big\vert\ 0.1\ M\ HClO_4,\ H_2O\ \Big\vert\ \begin{matrix} 10^{-3}\ M\ Fe(ClO_4)_3,\ 10^{-3}\ M\ Fe(ClO_4)_2 \\ 0.1\ M\ HClO_4 \end{matrix}\ \Big\vert\ \overset{2}{\underset{\text{Pt}}{}} $$

(2.17)

In view of the fact that electrode 2 is influenced by the oxidation–reduction system (2.1), while electrode 1 is affected by system (2.14), the energy of the Fermi level of the electrons differs between these solutions. At the interface between the solutions only the liquid junction potential (see p. 57) is formed which, however, is negligible because the solutions contain 0.1 M $HClO_4$ as indifferent electrolyte. Thus, we shall assume that the inner electrical potentials of both solutions are identical

$$\varphi(\text{solution 1}) = \varphi(\text{solution 2})$$

The inner potential of electrode 2 is given by equation (2.8) where, of course, $\varphi(\text{Pt}, 1)$ is replaced by $\varphi(\text{Pt}, 2)$,

$$\varphi(\text{solution } 2) = (\mu_{\text{Fe}^{2+}}^0 - \mu_{\text{Fe}^{3+}}^0 - \mu_{\text{e}}^0(\text{Pt}))/F$$
$$- (RT/F)\ln(a_{\text{Fe}^{3+}}/a_{\text{Fe}^{2+}}) + \varphi(\text{Pt}, 2) \qquad (2.19)$$

Equations (2.7) and (2.19) yield

$$\mu_{\text{e}}^0(\text{Pt}) - \varphi(\text{Pt}, 1) = (RT/2F)\ln p_{\text{H}_2} + (2.3RT/F)pH - F\varphi(\text{solution } 2) \qquad (2.20)$$

Combining equations (2.19) and (2.20) gives

$$\Delta V = \varphi(\text{Pt}, 2) - \varphi(\text{Pt}, 1)$$
$$\Delta V = (\mu_{\text{Fe}^{3+}}^0 - \mu_{\text{Fe}^{2+}}^0)/F + (RT/F)\ln(a_{\text{Fe}^{3+}}/a_{\text{Fe}^{2+}})$$
$$+ (2.3RT/F)pH + (RT/2F)\ln p_{\text{H}_2} \equiv E \qquad (2.21)$$

The difference in the electrical potentials of two electrodes that are in equilibrium with adjacent solutions in contact is called the Electromotive Force (EMF) and denoted E. The system of two electrodes with these properties is called a *galvanic cell*. It can be seen from the right-hand side of equation (2.21) that it contains only quantities that can be determined experimentally. Quantities that require special assumptions, like the inner electrical potentials of individual phases, are absent.

Galvanic Cells

In the last century, M. Faraday was the first to show that the EMF of a galvanic cell expresses the energy gained from a reaction that takes place in the cell. Faraday's idea was at that time rather nebulous (he did not know the proper definition of energy) but his intuition was excellent. This concept was refined by Bethellot, Gibbs, and Helmholtz and, thus, these scientists are responsible for the complete theory of the EMF of a galvanic cell. The right-hand side of equation (2.21) can readily be expressed in terms of equation (2.16) using chemical potentials

$$EF = -(\mu_{\text{Fe}^{2+}} + \mu_{\text{H}_3\text{O}^+} - \mu_{\text{Fe}^{3+}} - \tfrac{1}{2}\mu_{\text{H}_2} - \mu_{\text{H}_2} - \mu_{\text{H}_2\text{O}}) \qquad (2.22)$$

The expression in parentheses is the sum of the products of the chemical potentials and stoichiometric coefficients (see p. 65) which expresses the Gibbs energy change in the reaction

$$\text{Fe}^{3+} + \tfrac{1}{2}\text{H}_2 + \text{H}_2\text{O} = \text{Fe}^{2+} + \text{H}_3\text{O}^+ \qquad (2.23)$$

The numbers of mols of reactants and products is the same as the corresponding stoichiometric coefficients (the reaction occurs to *unit extent*). One mol of Fe^{3+} and one mol of H_2O react with one-half mol of H_2 (their stoichiometric coefficients are negative since they are reactants) yielding as products one mol of Fe^{2+} and one mol of H_3O^+ (with positive stoichiometric coefficients as defined for the reaction products).

A scheme of the cell (2.17) reveals that the materialization of reaction (2.23) is connected with a negative current flowing through the cell from right to left. The Fe^{3+} ions are reduced by the electrons coming from electrode 2 and the hydrogen molecules transfer electrons to electrode 1 (which is identical to a positive current flowing from left to right). The total electric charge consumed during current flow corresponds to one mol of electrons, consequently $1F$ is transferred through the cell from electrode 1 to electrode 2.

Reaction (2.23) will be accomplished in the cell, for example, so that a somewhat smaller voltage is connected with the electrodes of the cell than the EMF of the cell (or the cell is simply short-circuited across a large resistance). Then reaction (2.23) will take place spontaneously. Such a reaction is characterized by a negative Gibbs energy change ($\Delta G < 0$, see Appendix B) and the resulting ΔV is positive. Generally, if the cell reaction takes place spontaneously, the resulting EMF is positive.

The graphical scheme (2.17) can also be written in the reverse direction,

$$
\begin{array}{c|c|c|c|c}
1 & & & & 2 \\
\text{Pt} & 10^{-3} \text{ M Fe(ClO)}, \ 10^{-3} \text{ M Fe(ClO)}, & 0.1 \text{ M HClO}_4 & \text{H}_2 & \text{Pt} \\
& 0.1 \text{ M HClO}_4 & & &
\end{array}
$$

Obviously, the EMF of this cell, E', is equal to E multiplied by -1. When a positive current flows through the cell from left to right the reaction

$$\text{Fe}^{2+} + \text{H}_3\text{O}^+ = \text{Fe}^{3+} + \tfrac{1}{2}\text{H}_2 + \text{H}_2\text{O}$$

occurs in the cell but, of course, not spontaneously. It can be accomplished by imposing a voltage to the electrodes such that the resulting difference between the electrodes is larger in absolute value (i.e. 'more negative') than E'. Reaction (2.23) can also be written in a different way, for example in a twofold extent

$$2\,\text{Fe}^{3+} + \text{H}_2 + 2\,\text{H}_2\text{O} = 2\,\text{Fe}^{2+} + 2\,\text{H}_3\text{O}^+ \tag{2.24}$$

In order to realize this reaction, again to unit extent, a charge of $2F$ must flow through the cell (2.17). The Gibbs energy change resulting from reaction (2.24) is $\Delta G' = 2\Delta G$. These formal variations in writing the reaction must not influence the EMF value. Instead of equation (2.22)

$$2EF = -(2\mu_{\text{Fe}^{2+}} + 2\mu_{\text{H}_3\text{O}^+} - 2\mu_{\text{Fe}^{3+}} - \mu_{\text{H}_2} - 2\mu_{\text{H}_2\text{O}}) = -\Delta G'$$

is then written. Generally, if the flow of charge zF through the cell is necessary for accomplishing a cell reaction to unit extent, the EMF of that cell is given by the equation

$$E = -\Delta G/(zF) \tag{2.25}$$

Returning to cell (2.17), the standardization of the hydrogen electrode also requires a suitable pH of the solution and a suitable hydrogen gas pressure. *The standard hydrogen electrode* is immersed in a solution with unit hydronium ion activity ($pH = 0$) while the relative partial pressure of hydrogen is equal to unity.

The EMF of cell (2.17), where the standard hydrogen electrode is on the left-hand side

1				2
Pt	H_2	aqueous electrolyte	10^{-3} M Fe(ClO$_4$)$_3$	Pt
	$p_{H_2} = 1$	solution, $a_{H_3O^+} = 1$	10^{-3} M Fe(ClO$_4$)$_2$	
			$> a_{H_3O^+} = 1$	

is called the *electrode potential* of the electrode placed on the right-hand side in the graphical scheme of the cell, but only when the liquid junction between the solutions can be neglected or calculated, or if it does not exist at all. The electrode potential of the electrode on the right-hand side (often called the oxidation–reduction potential, or redox potential) is given by the *Nernst equation*

$$E_{Fe^{3+}/Fe^{2+}} = E^0_{Fe^{3+}/Fe^{2+}} + (RT/F)\ln(a_{Fe^{3+}}/a_{Fe^{2+}}) \qquad (2.26)$$

This relationship follows from equation (2.21) when $(\mu^0_{Fe^{3+}} - \mu^0_{Fe^{2+}})/F$ is set equal to $E^0_{Fe^{3+}/Fe^{2+}}$ and the pH and $\ln p_{H_2}$ are equal to zero. In the subscript of the symbol for the electrode potential the symbols for the oxidized and the reduced components of the oxidation–reduction system are indicated. With more complex reactions it is particularly recommended to write the whole reaction that takes place in the right-hand half of the cell after symbol E (the 'half-cell' reaction); thus, in the present case

$$E_{Fe^{3+}/Fe^{2+}} \equiv E(Fe^{3+} + e = Fe^{2+})$$

Quantity $E^0_{Fe^{3+}/Fe^{2+}}$ is termed the *standard electrode potential*. It characterizes the oxidizing or reducing ability of the component of oxidation–reduction systems. With more positive standard electrode potentials, the oxidized form of the system is a stronger oxidant and the reduced form is a weaker reductant. Similarly, with a more negative standard potential, the reduced component of the oxidation–reduction system is a stronger reductant and the oxidized form a weaker oxidant.

The values of standard electrode potentials help to explain the course of reactions (2.11) and (2.12). The standard electrode potential of the Fe^{3+}/Fe^{2+} system is $E^0_{Fe^{3+}/Fe^{2+}} = 0.771$ V, while for the V^{3+}/V^{2+} system $E^0_{V^{3+}/V^{2+}} = -0.20$ V. The Tl^{3+}/Tl^+ system is characterized by $E^0_{Tl^{3+}/Tl^+} = 1.25$ V, so that univalent thallium is not oxidized by trivalent iron but by tetravalent cerium ($E^0_{Ce^{4+}/Ce^{3+}} = 1.610$ V). It is important to note that a reaction will not occur in practice unless it has sufficient velocity. This condition is fulfilled with the majority of the examples of this paragraph while, for example, the oxidation of hydrogen by trivalent iron does not take place in practice, as already mentioned on p. 70.

In the above examples the course of a reaction is completely determined by the standard potential values because their differences are large. When the standard potentials are relatively close, the direction of the reaction depends on the values of the electrode potentials according to equation (2.26). Equilibrium is attained in the oxidation–reduction reaction when both the oxidation–reduction systems have the same redox potential.

The general form of equation (2.26) for the half-cell reaction

$$ox + ze = red$$

is

$$E_{ox/red} = E^0_{ox/red} + (RT/zF) \ln (a_{ox}/a_{red})$$
$$= E^0_{ox/red} + (2.3\ RT/zF) \log (a_{ox}/a_{red})$$

Coefficient $2.3RT/F$ equals 0.059 V at 25°C.

The electrode potential is not determined exclusively by the two components of the oxidation–reduction system present in the solution. A different case has already been mentioned for the hydrogen electrode, which is an example of a *gas electrode* (the reduced form is gaseous hydrogen) A *cationic electrode* is an electrode consisting of metal M immersed in a solution containing ions of the same metal M^{z+}. The half-cell reaction at this electrode is

$$M^{z+} + ze = M$$

The Nernst equation for a cationic electrode is

$$E_{M^{z+}/M} = E^0_{M^{z+}/M} + (RT/zF) \ln a_{M^{z+}} \qquad (2.28)$$

since the activity of the reduced form, i.e. of the electrode material, has been included as a constant term in $E^0_{M^{z+}/M}$. When the electrode is a dilute amalgam of the metal M contacting dissolved M^{z+} ions, the corresponding Nernst equation is

$$E_{M^{z+}/MHg_x} = E^0_{M^{z+}/MHg_x} + (RT/zF) \ln (a_{M^{z+}}/a_{MHg_x})$$

where a_{MHg_x} is the activity of the metal dissolved in mercury. Amalgam electrodes often have a considerable advantage compared with pure metal electrodes because the half-cell reactions are not hampered by other processes (with electrodes with rather negative standard potentials, hydrogen evolution is a complicating factor, as will be shown on p. 103).

It is not easy to obtain a perfectly functioning hydrogen electrode without observing several strict precautions: the hydrogen must be pure, the preparation of the electrode covered with platinum black is not simple, and it is necessary to prevent various impurities from reaching the neighbourhood of the electrode since they would 'poison' the catalytically active layer. In view of this, *electrodes of the second kind* are almost exclusively used as reference electrodes. These are metallic electrodes covered with a sparingly soluble salt of the corresponding metal ion with a suitable anion; a soluble salt of this anion with an alkali metal cation is present in the solution in contact with the electrode. The layer of the sparingly soluble salt is simultaneously in equilibrium with the metal and with the electrolyte solution.

The silver–silver chloride electrode consists of a piece of silver wire or a plate covered with a layer of silver chloride and immersed in an electrolyte solution containing chloride ions (e.g. KCl). The Nernst equation describing the behaviour of this electrode can be derived on the basis of a graphical diagram of a cell consisting of the standard hydrogen electrode and of the silver–silver chloride electrode (cf. scheme (2.25))

$$\text{Pt} \mid \text{H}_2 \mid \text{aqueous electrolyte} \mid \text{aqueous KCl} \mid \text{AgCl} \mid \text{Ag}$$
$$\text{solution, } a_{\text{H}_3\text{O}^+} = 1 \quad \text{solution,}$$
$$a_{\text{Cl}^-}$$

The cell reaction

$$\text{AgCl} + \tfrac{1}{2}\text{H}_2 + \text{H}_2\text{O} = \text{Ag} + \text{Cl}^- + \text{H}_3\text{O}^+$$

determines the EMF value

$$E = -(1/F)(\mu^0_{\text{Ag}} + \mu^0_{\text{Cl}^-} + RT\ln a_{\text{Cl}^-} - \mu^0_{\text{AgCl}}) \tag{2.29}$$

Thus, the Nernst equation becomes

$$E_{\text{AgCl}} \equiv E(\text{AgCl} + \text{e} = \text{Ag} + \text{Cl}^-)$$
$$= E^0_{\text{AgCl/Ag}} - (RT/F)\ln a_{\text{Cl}^-}$$

An analogous equation is valid for the calomel electrode which consists of a mercury electrode covered by calomel (HgCl) and a KCl solution. The saturated calomel electrode containing a saturated solution of potassium chloride is particularly important.

These electrodes are also used as electric potential probes in electrolyte where different points possess different values of potential. The electric potential difference between such electrodes indicates the electric potential difference between these points.

This approach has found frequent application in electrophysiology. In order to determine the electric potential difference, for example between two points in the extracellular liquid or between the inside and the outside of a cell (the membrane potential, see p. 133), micropipettes (Fig. 38) are very often used. These are

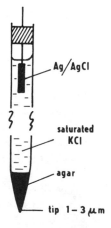

Fig. 38. A micropipette

silver–silver chloride electrodes filled with a saturated KCl solution immobilized with agar–agar gel. A capillary with a tiny tip (diameter 1–3 μm) connects the electrode to the point of measurement. Since the potassium ions often influence the electric properties of excitable cells (see p. 168), the saturated KCl solution is sometimes replaced with saline (0.9 % NaCl solution). In this case, of course, the liquid junction potential between the orifice of the capillary and the investigated solution cannot be neglected. On the other hand, a probe of this type correctly reflects the changes in electric potential.

More complicated electrode equilibria are found in organic oxidation–reduction systems which are often of quinoid character. The equilibrium between a quinone-type oxidized form and the hydroquinone-type reduced form is described by the equation

$$\text{quinone} + 2e = \text{hydroquinone} \tag{2.30}$$

Both forms, or at least the reduced form, are acids binding different numbers of hydrogen ions. Such an oxidation–reduction system is often characterized by the pH dependence of the redox potential, E_H^0, defined by the condition that the total concentration of all the components in the oxidized form (protonized to different degrees) is equal to the total concentration of all the components in the reduced form. Examples of this dependence are shown in Fig. 39. Curve 1 corresponds to the benzoquinone/benzohydroquinone system. In an aqueous solution the

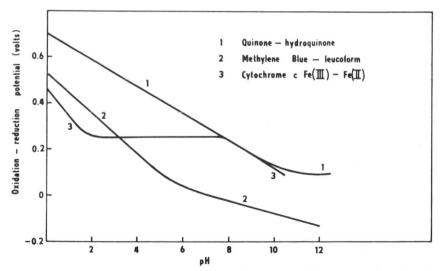

Fig. 39. The pH-dependence of organic oxidation–reduction potentials. The slope dE/dpH of the dependence for the quinone–hydroquinone system is −0.029 V and 0 V. The slope for the methylene blue–leucoform of methylene blue is −0.089 V for pH <6 and then becomes −0.029 V. The slope for the ferriferrocytochrome c is −0.118 V below pH 2, 0 V between pH 2 and 8, and −0.059 V at higher pH. The change from the oxidized form to the reduced form of this enzyme does not affect the acid-base properties of its protein component

quinone possesses neither acidic nor basic properties, while the hydroquinone is a diprotonic acid. Curve 2 shows the pH dependence of the redox potential of methylene blue. The acidobasic properties of this oxidation–reduction system consisting of a blue oxidized form and of a colourless reduced form (called the luecoform; Greek *leukos* = white) are demonstrated in the reaction scheme

pH < 6:

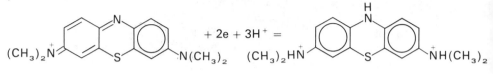

pH > 6:

Cytochrome c, an enzyme (Fig. 40), contains a porphyrin complex of iron as the prosthetic group with the oxidation–reduction properties described schematically by equation (2.3). The protein component of the enzyme is a polyelectrolyte. The protolytic reactions of the prosthetic group lead to the *pH* dependence of the redox potential (curve 3, Fig. 39).

Reaction (2.30) is, in fact, a combination of two reactions of successive binding of two electrons to the quinone molecule. First, the semiquinone is produced from the quinone

$$\text{quinone} + e = \text{semiquinone}$$

and then the hydroquinone is formed

$$\text{semiquinone} + e = \text{hydroquinone}$$

The semiquinone is often quite unstable so that its formation cannot be detected. The reduction of quinone is then expressed by equation (2.30). Sometimes, however, the semiquinone is relatively stable in spite of the fact that it is a free radical (it contains one unpaired electron).

The standard electrode potentials for various half-cell reactions are listed in order in Fig. 41. The higher the position of a cell reaction the stronger oxidant the oxidized form is, while the strength of the reductants increases towards the bottom of the scale.

A galvanic cell consists of two electrodes, half-cell reactions taking place at each of them. Let us imagine that a solution containing hydrogen ions of unit activity and the hydrogen electrode under standard hydrogen pressure is placed between the electrodes of the cell (see Fig. 42). In this way two galvanic cells are formed, one with the standard hydrogen electrode on the left-hand side and the other one with this electrode on the right-hand side. Obviously, the EMF of the

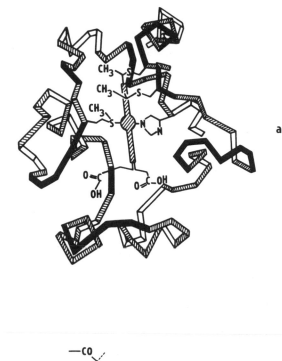

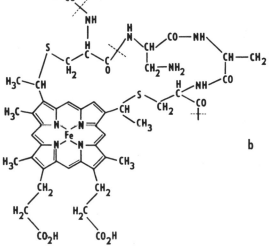

Fig. 40. Cytochrome c structure according to Dickerson and co-workers. In scheme (a) the individual aminoacids are represented by trapezoids. Shading indicates the distance from the observer. The rod in the centre is the lateral view of the heme, the porphyrin complex of iron. The sulphur atom of methionine and the nitrogen atom of histidine are located in the two remaining free coordination sites of the heme. The bounding of the protein chain to the upper part of the heme is shown in scheme (b)

80

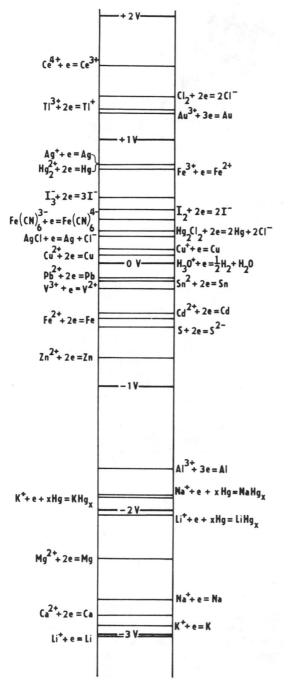

Fig. 41. Standard potentials of inorganic systems

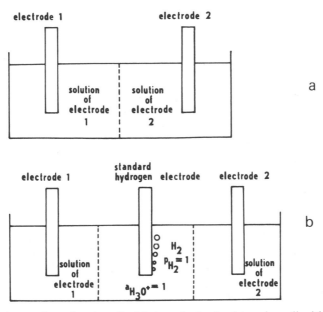

Fig. 42. A scheme of a galvanic cell with two electrodes (a) and a cell with an inserted standard hydrogen electrode (b)

first cell is equal to the electrode potential of this cell E (right-hand electrode) and the EMF of the second cell equals the electrode potential of the other electrode E (left-hand electrode) multiplied by -1. The overall EMF of the original cell (the fact that the standard hydrogen electrode has been connected to the cell has not changed the conditions in this cell) is consequently given by the difference between the potential of the right-hand electrode and of the left-hand electrode

$$E = E(\text{right-hand electrode}) - E(\text{left-hand electrode}) \qquad (2.31)$$

In this way the EMFs of galvanic cells consisting of arbitrary electrodes can be calculated, provided that the potentials of the liquid junctions between the electrodes (if they exist at all) are suitably eliminated. The result of this calculation will be in agreement with the EMF actually measured if the condition of the *reversibility* of the processes occurring in the cell is fulfilled. Suppose that a voltage which is slightly higher than the EMF of a cell is imposed on the electrodes of the cell. An electrical current will flow through the outer circuit and a cell reaction will begin in a certain direction. When a smaller voltage than the EMF is imposed, the direction of the current and of the cell reaction will be opposite. If the cell reaction is identical in both cases and differs only in its direction and, at the same time, it is identical with the cell reaction connected with the overall chemical transformation in the cell, this galvanic cell is called *reversible*.

Most galvanic cells, however, do not fulfil this condition. Because of the sluggishness of the overall cell reactions, the EMF is determined by partial reaction steps which are sometimes not even reversible (i.e. after current reversal another reaction takes place in the cell, etc.). For example, in the oxygen–hydrogen cell (Grove's cell)

$$Ni \,|\, H_2 \,|\, KOH,\, H_2O \,|\, O_2 \,|\, Ag$$

the expected cell reaction

$$O_2 + 2\,H_2 = 2\,H_2O$$

does not occur under current drain with an EMF of 1.23 V. The reversible half-cell reaction (2.14) occurs at the nickel electrode but at the right-hand electrode oxygen is partially reduced to the hydrogen peroxide anion

$$O_2 + H_2O + 2e = HO_2^- + OH^-$$

(cf. p. 102). The hydrogen peroxide formed is then decomposed on the electrode by a set of reactions to water and oxygen (see p. 103). When the current is reversed, water is not oxidized to oxygen but oxidation of the electrode takes place

$$2\,Ag + 2\,OH^- = Ag_2O + H_2O$$

These processes result in an EMF of less than 1 V, which is much less than expected from the cell reaction.

On the contrary, when the lead storage battery

$$Pb \,|\, PbSO_4 \,|\, H_2SO_4,\, H_2O \,|\, PbO_2 \,|\, Pb \qquad (2.32)$$

is discharged (external voltage lower than the EMF) the cell reaction

$$PbO_2 + Pb + 2\,H_2SO_4 = 2\,PbSO_4 + 2\,H_2O \qquad (2.33)$$

takes place. When the lead battery is charged, the same cell reaction commences in the opposite direction. This cell reaction corresponds to the overall chemical transformation in the cell during charging and discharging. Therefore, the lead accumulator is an example of a reversible galvanic cell.

Consider once again the experiments described on pp. 56, 57 and 58. The present notation will correspond to the cells

$$Ag \,|\, AgCl \,|\, 10^{-4}M\ HCl \,|\, 10^{-3}M\ HCl \,|\, AgCl \,|\, Ag \qquad (2.34)$$

and

$$Ag \,\Big|\, AgCl \,\Big|\, \begin{matrix} 10^{-4}\ M\ HCl \\ 0.1\ M\ KNO_3 \end{matrix} \,\Big|\, \begin{matrix} 10^{-3}\ M\ HCl \\ 0.1\ M\ KNO_3 \end{matrix} \,\Big|\, AgCl \,\Big|\, Ag \qquad (2.35)$$

Cells consisting of identical electrodes and differing in the concentration of adjacent electrolytes are termed *concentration cells*.

The measured values of the EMFs of these cells at 25°C are -21 mV for cell (2.34) and -59 mV for cell (2.35). In the former case a liquid junction potential is present in the cell, termed a *concentration cell with transport* (transference), so that the EMF is obtained by extending equation (2.31) by the liquid junction potential term

$$E = E(\text{right-hand electrode}) - E(\text{left-hand electrode}) + \Delta\varphi_L$$

(it should be borne in mind that $\Delta\psi_L$ is the electrical potential difference of the right-hand solution minus the left-hand solution). Thus, if the electrode responds to the ions I^{z_i} the EMF of the concentration cell is given by the equation

$$\begin{aligned} E &= (RT/z_iF)\ln\left[a_i(2)/a_i(1)\right] + \Delta\varphi_L \\ &= (RT/z_iF)\ln\left[c_i(2)/c_i(1)\right] \\ &\quad + (RT/z_iF)\ln\left[\gamma_i(2)/\gamma_i(1)\right] + \Delta\varphi_L \end{aligned} \tag{2.36}$$

As, in the present case, we have $z_i = 1$, $c_i(2) = 10^{-3}\,\text{mol/dm}^3$, $c_i(1) = 10^{-4}$ mol/dm^3, $\gamma_i(2)/\gamma_i(1) \approx 1$, and $\Delta\varphi_L = 38\,\text{mV}$ (p. 57) we obtain for the EMF the value of $-21\,\text{mV}$, in coincidence with the experiment. In the case of the *concentration cell without transport* (2.35), $\Delta\varphi_L \approx 0$ and the EMF has the value $-59\,\text{mV}$.

The basic equation (2.25) determines the upper limit for the transformation of the energy of a chemical reaction to the useful (electrical) energy with maximum efficiency. The Gibbs energy change is the measure of the exploitability of the energy of a chemical reaction regardless of the possibility of achieving this degree of exploitation.

When producing electrical energy in thermal power plants we also employ a chemical reaction, namely the combustion of carbonaceous fuels like coal, oil, or natural gas, for example,

$$C + O_2 = CO_2 \tag{2.37}$$

This reaction is connected with the release of the reaction heat (negative reaction enthalpy—see Appendix B)—$\Delta H = 393\,\text{kJ}$. The steam formed by feeding this amount of heat to the boiler drives the turbine, cools down at the same time and, at last, supplies heat to a heat exchanger and condenses to water. The utilization of the gained energy 393 kJ per 1 mol of carbon is limited by the *efficiency of the heat engine*. The heat engines include the steam engine, the steam turbine, the internal combustion engines, the magnetohydrodynamic power generator, the thermionic energy exchanger, etc. These all are devices that transform heat to other energy forms, mechanical or electrical. The heat is gained from a thermal source with a definite temperature while part of it is supplied to a heat exchanger with a lower temperature. The efficiency of energy transformation η is given by the ratio of the gained useful work $-W$ and of the heat supplied from the thermal source $-Q$

$$\eta = W/Q = (T_1 - T_2)/T_1$$

where T_1 is the temperature of the thermal source and T_2 is that of the cooler heat exchanger (in Kelvins). For high quality power stations $\eta \approx 0.45$.

The galvanic cell is not a thermal engine. It directly converts the maximum share of the energy of a chemical reaction (i.e. the Gibbs energy) to useful, electrical work according to equation (2.25). In order to achieve maximum utilization of the energy the electrical potential difference between the electrodes $\Delta V'$ under which the electrical current is drawn from the cell must not differ markedly from the EMF (i.e. the difference $E - \Delta V'$ must be small). In addition, the conditions for reversibility of the cell must be fulfilled.

The Gibbs energy change $\Delta G < 0$ and the heat gained from a chemical reaction $\Delta H < 0$ are related by the equation (see Appendix B)

$$-\Delta G = -\Delta H + T\Delta S$$

where ΔS is the entropy change during the reaction. ΔS may sometimes be positive so that the efficiency of the energy transformation

$$\eta = \Delta G/\Delta H$$

is larger than 1. As the product $T\Delta S$ is the heat supplied to the cell from its surroundings when the cell reaction takes place close to equilibrium, even this amount of heat is transformed to useful work. The lead accumulator is an example of this situation. The entropy change accompanying reaction (2.33) is $+42.6 \, J.K^{-1}.mol^{-1}$, so that the heat from the surroundings also contributes somewhat to the transformation of the chemical energy to electrical energy when discharging the battery. This energy balance is in no way contrary to the Second Law of Thermodynamics, as when charging the accumulator an equal amount of heat is transferred to the surroundings.

The use of galvanic cells (or electrochemical power sources, as they are sometimes called in practical applications) began quite early in the development of electricity theory as well as of chemistry. Textbooks usually state that the first galvanic cell was constructed by Alessandro Volta, professor at the University of Pavia, in 1800. Volta's 'pile' was a column of zinc and copper discs separated by felt wetted with dilute sulphuric acid. However, the first cell was probably built by Alexander von Humboldt three years earlier. For the most part of the nineteenth century, Volta's pile (and other galvanic cells in the following decades) was used as the only source of 'galvanic electricity', i.e. of electric current, in contrast to 'voltaic electricity' obtained by various types of electric machines (see Fig. 43). At the end of the century the dynamo invented by W. von Siemens in 1866 finished with the monopoly of electrochemical power sources in the field of electricity generation.

From the standpoint of the origin of chemical energy in practical galvanic cells they can be divided into three groups.[1]

Primary cells[2] already contain the reagents of their energy producing cell reactions when they are assembled. The most frequently used member of this group is the Leclanché or manganese dioxide cell (EMF = 1.55 V). It was invented by the French engineer Georges-Lionel Leclanché (1838–1882) in 1866. It contains a carbon electrode covered with a layer of a mixture of manganese dioxide (pyrolusite) and soot suspended in a solution of ammonium chloride. The other electrode is made of zinc, which is usually amalgamated. The supply of

Fig. 43. A gigantic voltaic battery placed by H. Davy in the cellar of the Royal Institution of Great Britain. According to an early nineteenth-century engraving drawn by J.D.

electrical current is connected with the cell reaction

$$MnO_2 + Zn + 8\,NH_4^+ + 2\,H_2O = Mn(NH_3)_4^{2+} + Zn(NH_3)_4^{2+} + 4\,H_3O^+$$

The usual version of the Leclanché cell, called the dry cell (Fig. 44), was invented much later. The cell is not actually dry, however, since it contains a solution of ammonium chloride thickened with wheat flour. This cell represents 90% of global production of primary cells; production in the U.S. equals more than three billion cells per year.

Another important primary cell is the Weston cell

$$CdHg_x(12.5\,wt\,\%\,Cd)\,\bigl|\,3\,CdSO_4 \cdot 8\,H_2O,\,Hg_2SO_4\,\bigr|\,Hg$$
$$\text{saturated} \quad\quad \text{solid}$$
$$\text{aqueous}$$
$$\text{solution}$$

with the EMF = 1.01807 V at 25°C. The Weston cell represents a standard voltage because, when carefully prepared, it retains the required EMF value with great accuracy. The temperature coefficient of the cell is small and a constant composition can be maintained for a long time.

Similar cells used as sources for miniature electrical appliances (hearing aids, pacemakers, etc.) is the mercury cell which is similar to the Weston cell with cadmium replaced by zinc or the indium cell with indium instead of cadmium. 'Superdry' cells have a solid electrolyte, for example the cell

$$Ag\,|\,AgBr\,|\,CuBr\,|\,Cu.$$

The solid electrolytes AgBr and CuBr have very low conductivity, so that the cell only acts as a long-life voltage source.

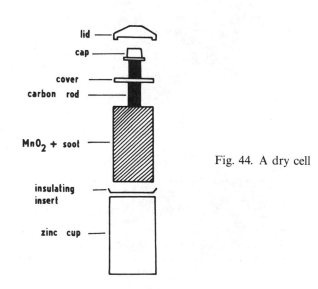

Fig. 44. A dry cell

Secondary cells[3] (accumulators, storage batteries) are galvanic cells where the power supplying reagents are formed only when the cell is charged from an external source of electric current. Thus, they serve for accumulation of electric energy. The most important (and the oldest) secondary cell is the lead accumulator. The originators of this storage battery are usually given as H. Sinsteden (1854) and G. Planté (1859) but it may be concluded from a much older Juvenile Lecture by M. Faraday that he was acquainted with the principle of the reaction in this cell. A virtually working lead accumulator was probably first constructed by Planté in 1860. Lead accumulators are used primarily as batteries in cars, as power sources for railway signals, for runabouts (the only 'electromobiles' really used in practice), etc.

The charging/discharging twin process is called the accumulator cycle. A starter battery must last for 400 cycles. A disadvantage of the lead accumulator is its large weight per unit of stored charge. Better weight parameters are exhibited by nickel–cadmium and silver–zinc batteries which are used in portable appliances (for example, radios, TV sets) and aircraft.

The third type of galvanic cells are *fuel cells*.[4] Here, the cell reagents are continuously supplied to the cell during current drain. An ancient dream of electrochemists is the realization of the reaction of fuel burning (2.37) as a practicable cell reaction. This would mean that coal would be oxidized at the negative electrode of the cell while oxygen would be reduced at the positive electrode. An attempt was made to construct such a cell, for example by the German chemists Wilhelm Ostwald and Fritz Haber (both Nobel Prize winners, but not for fuel cell discovery). At elevated temperatures, direct coal oxidation in a cell is possible, particularly when a fused electrolyte is used, but such cells are unstable and unsuitable for a long operation. Somewhat better results were achieved using carbon monoxide as a cell fuel. Oxidation occurs in a strongly acidic solution with platinum or palladium anodes. In another version a solid electrolyte, the Nernst mass $(85\% ZrO_2 + 15\% Y_2O_3)$ is used and carbon monoxide is oxidized and oxygen reduced at platinum electrodes.

Compounds like methanol, formic acid, oxalic acid, etc. show better performance as fuels. At higher temperatures in strongly acidic solution even simple hydrocarbons like methane, ethane, and ethylene are also oxidized. However, none of these systems has been used in practical applications.

The only exception is the oxygen–hydrogen cell mentioned on p. 82 which also belongs among fuel cells because gaseous hydrogen is the fuel. This cell does not transform chemical energy into electrical energy ideally but has the advantage of low weight per amount of energy supplied. This low weight was crucial in employing the oxygen–hydrogen fuel cells as the electric energy sources in space vehicles. The American space ship Gemini (Latin *gemini* = twins, as it carried two astronauts on board) contained this cell equipped with platinum electrodes and an ion-exchanging membrane (see p. 132) as cell electrolyte (a thin film of liquid electrolyte had to be present at the surface of each electrode contacting the membrane). The membrane simultaneously prevents the reacting gases from reaching the electrodes which they do not correspond to, i.e. the hydrogen to the oxygen electrode and vice versa. In the subsequent Apollo missions (which culminated in the lunar landing of the astronauts in 1969) oxygen–hydrogen cells with a liquid electrolyte were used (Fig. 45).

In addition to chemical energy, light energy can be transformed into electrical energy in electrochemical devices. These *electrochemical photovoltaic cells* will be dealt with in the section on photoelectrochemistry (p. 115).

References

1. V. S. Bagotzky and A. M. Skundin, *Chemical Power Sources*, Academic Press, New York, 1980.
2. G. W. Heise and N. C. Cahoon, Eds., *The Primary Battery*, John Wiley & Sons, New York, 1968.
3. H. Bode, *Lead–acid Batteries*, John Wiley & Sons, New York, 1977.
   S. U. Falk and A. J. Salkind, *Alkaline Storage Batteries*, John Wiley & Sons, New York, 1969.
4. W. Vielstich, *Fuel Cells: Modern Processes for the Electrochemical Production of Energy*, Wiley–Interscience, New York, 1970.
   M. W. Breiter, *Electrochemical Processes in Fuel Cells*, Springer, New York, 1969.

Voltammetry

Let us return to the experiment described on p. 62. Here, the dependence of the electric current on the electrode potential during current flow (the *polarization curve* or the *voltammogram*) is of interest. The experimental arrangement shown in Fig. 33 has, however, a definite drawback. The ohmic potential drop formed between electrodes 1 and 2 (see p. 64) cannot usually be neglected. Thus, in order to record the required dependence, it must be eliminated in a suitable way. When the resistance of the whole electrolysis system (i.e. the resistance between the metallic leads to the electrodes) R is known, than quantity IR can be simply subtracted from the voltage imposed on the electrodes. As a rule, however, a three-electrode arrangement* is used (see Fig. 46). The reference electrode (either made of the same material and present in the same solution as the polarizable electrode or a calomel or silver–silver chloride electrode with a salt bridge) is attached through the thin glass tube (the *Luggin capillary*) as close as possible to the polarizable (*working* or *indicator*) electrode. Under the flow of electric current between the indicator (polarizable) electrode 2 and electrode 1 (called the *auxilliary* or *current supplying electrode*) the difference in electrical potential between the indicator and the reference electrode, $E - E_{ref}$, the *polarization* of the electrode, is measured potentiometrically. Since the potential of the reference electrode E_{ref} is known, this measurement yields the polarization curve. The whole procedure is very frequently carried out automatically using an electronic or computer controlled *potentiostat*. This device maintains continuous control of the voltage applied between the indicator and the reference electrode. This

---

* In polarography a two-electrode system is often used since the IR-drop can be neglected in view of rather low currents.

Fig. 45(a) The General Electric fuel cells from the space capsule of Gemini V with 3 groups of 32 fuel cells in a pressurized container.
(b) The 500 W fuel cell system of Apollo Project, produced by Pratt & Whitney Corp. The fuel cells are ordered according to vertical axis in the lower part of the figure, the auxilliary devices are shown in the upper part of the figure

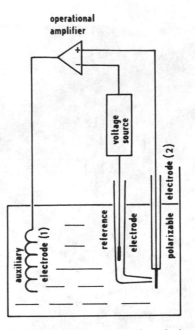

Fig. 46. Potentiostatic electrode polarization

voltage, fed from a suitable source, may be constant or may follow a required time dependence program (linear increase or decrease, triangular function, step-wise potential change, etc.).

There are two basically different voltammogram forms. When the solution is stirred (for example, when a rotating or vibrating electrode is used, see p. 119) or when the dropping mercury electrode is used (see p. 118) the common voltammogram (or polarogram) form is obtained (see Fig. 47a). The S-shaped part of the curve (the 'wave') shows that an electroactive substance (a depolarizer) present in the electrolyte solution is oxidized or reduced at the indicator electrode. The value of the current corresponding to the horizontal part of the curve is called the *limiting current*. This is often the analytically significant quantity and is usually directly proportional to the concentration of the electroactive species. The position of the wave on the potential scale is characterized by the *half-wave potential*, $E_{1/2}$. When a constant potential is applied to an unstirred (stationary) indicator electrode a decrease of the initially high current (a chronoamperometric curve) is sometimes observed, as already pointed out on p. 63. The decrease is usually due to the diffusion of the electroactive species to the electrode (see p. 64). A *peak voltammogram* is obtained when a linearly increasing (or decreasing) voltage (a 'potential sweep') is imposed on the indicator electrode (see Fig. 47b). In addition to single-sweep voltammetry, *cyclic voltammetry*, based on triangular voltage pulses applied to the indicator electrode, is frequently used. The voltammogram shown in Fig. 47c is then obtained. On the ascending side of the potential triangle an electroactive substance is oxidized and the oxidation product is re-reduced on the descending side.

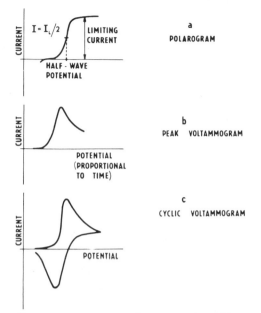

Fig. 47. Typical voltammograms. a, polarogram (= voltammogram); b, peak voltammogram; c, cyclic voltammogram

However, the nature of the electrode reaction often prevents the second current peak from appearing.[1]

References

1. P. Delahay, *New Instrumental Methods in Electrochemistry*, Interscience, New York, 1954.
R. S. Nicholson and I. Shain, *Anal. Chem.* **36** (1964) 706.
A. J. Bard and L. R. Faulkner, *Electrochemical Methods: Fundamentals and Applications*, John Wiley and Sons, Chichester, 1980.
D. T. Sawyer and J. L. Roberts, *Experimental Electrochemistry for Chemists*, Wiley-Interscience, New York, 1974.
E. Yeager and A. J. Salkind (Eds.), *Techniques of Electrochemistry*, Vol. 1–3, Wiley–Interscience, New York, 1972–1978.
E. Gileadi, E. Kirowa-Eisner and J. Penciner, *Interfacial Electrochemistry—an Experimental Approach*, Addison-Wesley, Reading, Mass, 1975.

Electrode Reactions

Consider the electrode reactions taking place in the system 10^{-3} M $Fe(ClO_4)_3$, 10^{-3} M $Fe(ClO_4)_2$ in 0.1 M $HClO_4$ (cf. p. 62). Work in a stirred solution will be achieved by employing a rotating electrode made of a platinum disc fixed in a Teflon tube. The axis of the tube is connected to a small geared electric motor so that the electrode can rotate with different velocities around the axis perpendicular to the surface of the disc. The reference electrode is a platinum wire in the solution of the same composition as the electrolysed solution. The stirring of the electrolyte solution prevents the decrease of the current with time described in connection with the experiment on p. 63.

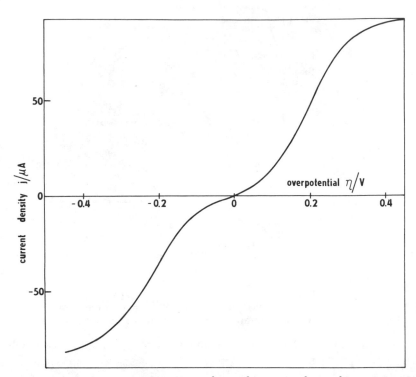

Fig. 48. Voltammogram of a solution of 10^{-3} M Fe^{3+} and 10^{-3} M Fe^{2+} in 0.5 M $HClO_4$ obtained on a rotating disc electrode. Rotation speed $N = 100\,\text{min}^{-1}$. According to J. Weber, Z. Samec, and V. Mareček

Generally, when the reference electrode is in the same solution as the working electrode, except in the currentless state, the difference between the electrode potentials is called the overpotential ('overvoltage') η. The voltammogram recorded using this electrode system is shown in Fig. 48 in coordinates current density j (electric current divided by electrode area) $- \eta$. A clearer insight into this dependence will be obtained when the voltammogram is plotted in coordinates $\log |j| - \eta$ (Fig. 49). It may be concluded from the log-linear portions of this dependence that the current density is an exponential function of the overpotential η. This feature is found for the cathodic as well as for the anodic wave on the voltammogram. More careful analysis of this dependence reveals that the current density is governed in the corresponding overpotential ranges by the equation for the anodic current density

$$j_a = j_0 \exp\left[(1 - \alpha)\, F\eta/RT\right] \tag{2.38}$$

and for the cathodic current density

$$j_c = -j_0 \exp\left(-\alpha F\eta/RT\right) \tag{2.39}$$

where F is the Faraday constant, R the gas constant, T the absolute temperature, α the charge transfer coefficient characterizing any electrode reaction, and j_0 the exchange current density. The latter quantity will be explained subsequently.

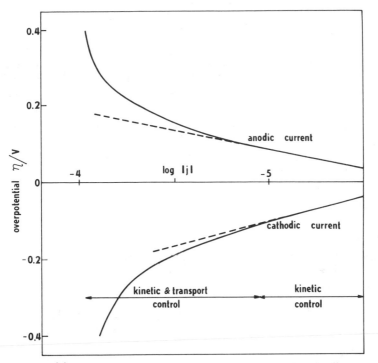

Fig. 49. The log $|j| - \eta$ plot for the voltammogram in Fig. 48. The linear portions correspond to the current controlled by the electrode reaction rate. The slopes are $2.3RT/(1 - \alpha)F$ (anodic current) and $-2.3RT/\alpha F$ (cathodic current) giving $\alpha = 0.47$. This is only an approximation, as shown in Appendix C

These two dependences are drawn as dashed lines in Fig. 49. As already pointed out on p. 64, the rate of an electrode reaction depends both on the rate of the oxidation (anodic) reaction and on the rate of the reduction (cathodic) reaction. The electrode reaction rate v is equivalent to the current density j, $j = Fv$.

Obviously, the rates of the oxidation and of the reduction reaction according to equation (2.2) are equivalent to the anodic and cathodic current densities, so that the overall current density is obtained by combining equations (2.38) and (2.39)

$$j = j_0\{\exp\left[(1 - \alpha)F\eta/RT - \exp\left[-\alpha F\eta/RT\right]\right\} \qquad (2.40)$$

A value of $\eta = 0$ (i.e. the system is in equilibrium) does not mean that the electrode reaction stops. According to equations (2.38) and (2.39) the anodic and the cathodic current densities are then $j_a = j_0$ and $j_c = -j_0$. Consequently, the anodic and the cathodic current densities at equilibrium are equal to the positive or the negative value of the exchange current density which cancel each other out. The existence of the exchange current can be directly proved by labelling one of the reacting component with a radioactive tracer. This is comparatively easy to achieve at amalgam electrodes (p. 75).

Equation (2.40) can be used to describe the electrode reaction rate in terms of the exchange current density j_0, the charge transfer coefficient α, and also the equilibrium potential E_{eq}. Changing the concentration of Fe^{3+} and Fe^{2+} reveals that j_0 depends on these concentrations while α remains constant (at least in practice). Thus, these two quantities, E_{eq} and j_0 are functions of the concentrations of the oxidized and of the reduced forms of the oxidation–reduction system. It would be preferable to characterize the electrode reaction rate by constants alone. Three partly new parameters can be introduced: the standard rate constant of the electrode reaction k^0, the standard potential of the corresponding half-cell reaction $E^0_{Fe^{3+}/Fe^{2+}}$, and the already known charge transfer coefficient, α. In equation (2.2) it was assumed that the rates of the electrode reaction depend linearly on $c_{Fe^{3+}}$ and $c_{Fe^{2+}}$, so that the rate constants k_{ox} and k_{red} are independent of $c_{Fe^{3+}}$ and $c_{Fe^{2+}}$. In order to obtain a suitable relationship for j_0, $E - E^0_{Fe^{3+}/Fe^{2+}} + (RT/F)\ln(c_{Fe^{3+}/Fe^{2+}})$ can be substituted for η in equation (2.40) (assuming $\gamma_{Fe^{3+}} = \gamma_{Fe^{2+}}$, cf. (2.26)). Thus, the electrode reaction rate equation will be obtained in form (2.2) only when

$$j_0 = Fk^0 c^{1-\alpha}_{Fe^{3+}} c^{\alpha}_{Fe^{2+}} \tag{2.41}$$

The basic electrochemical kinetic equation (or the Butler–Volmer equation), which is the final result of this calculation, has the following form for the Fe^{3+}/Fe^{2+} system

$$j = Fk^0\{c_{Fe^{2+}} \exp[1-\alpha)F(E - E^0_{Fe^{3+}/Fe^{2+}})/(RT)] \\ - c_{Fe^{3+}} \exp[-\alpha F(E - E^0_{Fe^{3+}/Fe^{2+}})/(RT)]\} \tag{2.42}$$

Consequently, the rate constants from equation (2.2) are given by the relationships

$$k_{ox} = Fk^0 \exp[(1-\alpha)F(E - E^0_{Fe^{3+}/Fe^{2+}})/(RT)]$$
$$k_{red} = Fk^0 \exp[-\alpha F(E - E^0_{Fe^{3+}/Fe^{2+}})/(RT)]$$

The standard electrode reaction rate constant here is $k^0 = 1.72 \times 10^{-5}\,\text{cm}\cdot\text{s}^{-1}$. A more general form of equation (2.42) for the electrode reaction red = ox + ze has the form

$$j = zFk^0\{c_{red} \exp[1-\alpha)zF(E - E^0_{ox/red})/(RT) \\ - c_{ox} \exp[-\alpha zF(E - E^0_{ox/red})/(RT)]\} \tag{2.43}$$

Both α and k^0 which characterize a given electrode reaction strongly depend on the composition of the solution. Several examples of these parameters together with further properties of the systems under study are listed in Table 8.[1]

It should be noted that, setting $j = 0$ (equilibrium state) in equation (2.43) would yield the Nernst equation (2.26).

We will now consider the transition of the initial exponential form of the voltammogram to the S-shaped form with the formation of limiting currents. The acceleration of an electrode reaction by increasing (or decreasing) the electrode potential results in a depletion of the reacting species in the vicinity of the electrode, while the concentration of the reaction product increases. The stirring of the electrolyte solution eliminates these concentration changes to a certain degree, but the rate of the transport process, termed *convective diffusion*, contributes to the rate of the overall process. This results in a lower absolute value

Table 8   Basic parameters of selected electrode reactions

Reaction	Electrolyte	Electrode	°C	$k°$ cm s^{-1}	α
$Cd^{2+} + 2e = CdHg_x$	1 M NaClO$_4$	Hg	25	0.46	0.32
$Ce^{4+} + e = Ce^{3+}$	1 M H$_2$SO$_4$	Pt	25	3.7×10^{-4}	0.21
$Hg_2^{2+} + 2e = 2 Hg$	45 % HClO$_4$	Hg[a]	-38.8	0.05	0.3
$Hg_2^{2+} + 2e = 2 Hg$	45 % HClO$_4$	Hg[b]	-38.8	0.043	0.3
$Zn^{2+} + 2e = ZnHg_x$	0.5 M NaClO$_4$	Hg	25	2×10^{-4}	0.26

[a] Liquid mercury.
[b] Solid mercury.

of the current than when the electrode reaction is the only current-determining process. When the electrode reaction rate is rather rapid, the concentrations of the species entering the electrode reaction fall to zero and a *limiting* current is observed. This kind of a current is controlled solely by diffusion with no contribution from the rate of the electrode reaction.

When the rate of stirring is increased, the convective diffusion is still able to maintain a constant concentration of the electroactive species at higher current densities than at lower rates of stirring. Under these conditions the limiting current is higher. In general, the rate of a process controlled, at least in part, by transport (for example, by convective diffusion) increases when the transport rate increases (for example, because of more intense stirring).

To clarify this situation, the concentration distributions of both the electroactive species Fe^{3+} and Fe^{2+} in the neighbourhood of the electrode (Fig. 50) must be considered. Curves 1 illustrate the concentration distribution for a low rate of electrode rotation, while curves 2 correspond to a higher rotation rate. The concentration distribution of the transported species is characterized by a concentration change from its value in the bulk of the electrolyte c_0 to the values c_{red}^* and c_{ox}^* directly at the electrode surface. These values depend on the overpotential as well as on the transport rate. For convective diffusion the region of decreased or increased concentration in the vicinity of the electrode is called the *Nernst layer*. Its effective thickness, δ_{red} or δ_{ox}, corresponding to Fe^{2+} and Fe^{3+}, respectively, is defined in the same way as the effective thickness of the diffusion layer (cf. Figs. 24 and 50).

For convective diffusion the transport rate in a definite direction depends on the diffusion as well as on the streaming of the medium. However, for transport perpendicularly towards the electrode, the diffusion flux at the electrode surface plays the only role because the electrode is impermeable for the solution. The diffusion flux is equivalent to the electric current according to the Faraday Law (1.30). Concentration gradients at the electrode surface are described by the equations (cf. Fig. 50)

$$(dc_{red}/dx)_{x=0} = (c_0 - c_{red}^*)/\delta_{red}$$
$$(dc_{ox}/dx)_{x=0} = -(c_0 - c_{ox}^*)/\delta_{ox} \tag{2.44}$$

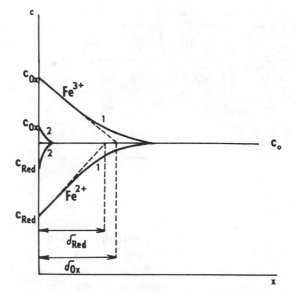

Fig. 50. Concentration distribution in the vicinity of the rotating electrode for low (1) and high (2) rotation speeds

By combining Fick's Law (1.26), the Faraday Law (1.30), and relationships (2.44), the equation for the current density is obtained

$$j = zFJ_{red} = zFD_{red}(dc_{red}/dx)_{x=0} = zFD_{red}(c_0 - c_{red}^*)/\delta_{red}$$
$$= -zFJ_{ox} = -zFD_{ox}(dc_{ox}/dx)_{x=0} = -zFD_{ox}(c_0 - c_{ox}^*)/\delta_{ox} \qquad (2.45)$$

where z is the charge number of the electrode reaction (zF is the charge necessary for an electrode reaction to take place to unit extent; consequently, for the reaction $Fe^{2+} = Fe^{3+} + e$, $z = +1$, for the reaction $Fe^{3+} + e = Fe^{2+}$, $z = -1$, etc.), J_{red} and J_{ox} are the fluxes of Fe^{2+} and Fe^{3+}, and D_{red} and D_{ox} are the diffusion coefficients of these substances, respectively.

For the anodic limiting current density $j_{d,a}$, $c_{red}^* = 0$, so that

$$j_{d,a} = zFD_{red}c_0/\delta_{red} \qquad (2.46)$$

For the cathodic limiting current, $j_{d,c}$, an analogous relationship

$$j_{d,c} = -zFD_{ox}c_0/\delta_{ox} \qquad (2.47)$$

is obtained. As equation (2.43) describes the conditions in the vicinity of the electrode, the concentrations of the electroactive substances at the electrode surface c_{ox}^* and c_{red}^* should be used in place of the bulk concentrations. Thus, combining the modified equation (2.43) and relationships (2.45) yields a description of the voltammogram in the terms of the electrode reaction and transport rates. However, as the resulting general equation is unfortunately rather clumsy, only two special cases will be considered. When the standard electrode reaction rate constant is low, for example for the ferric/ferrous system under our experimental conditions, a large overpotential must be applied in order to produce a measurable anodic or cathodic current. Thus, for the voltammogram shown in Fig. 48, the current is controlled either by the first term on the right-hand side in equation (2.42) and becomes anodic, or by the second term and is then cathodic. This simplifies things quite considerably, since using equation (2.43) without the second right-hand side term, equations (2.45) and (2.46) yield the simple equation for the irreversible anodic voltammogram (or 'irreversible wave')

$$E = E^0_{\text{ox/red}} - \frac{RT}{\alpha zF} \ln \frac{k^0 \delta_{\text{red}}}{D_{\text{red}}} + \frac{RT}{\alpha zF} \ln \frac{j_{d,a} - j}{j} \qquad (2.48)$$

For the half-wave potential ($j = \frac{1}{2} j_{d,a}$) it holds that

$$E_{1/2} = E^0_{\text{ox/red}} - \frac{RT}{\alpha zF} \ln \frac{k^0 \delta_{\text{red}}}{D_{\text{red}}} \qquad (2.49)$$

This quantity depends both on the thermodynamic (E^0) and kinetic (k^0) properties of the reacting system, as well as on the transport rate ($\delta_{\text{red}}/D_{\text{red}}$).

When, on the other hand, the standard rate constant of the electrode reaction is very high, both the terms on the right-hand side of equation (2.44) not only simultaneously influence the current but can even preserve the equilibrium between the oxidized and the reduced component according to the Nernst equation (2.26)—with c^*_{ox} and c^*_{red} instead of $a_{\text{Fe}^{3+}}$ and $a_{\text{Fe}^{2+}}$. Then, equations (2.27), (2.45), (2.46), and (2.47) yield the equation for the voltammogram of the 'reversible wave'

$$E = E^0_{\text{ox,red}} - \frac{RT}{zF} \ln \frac{D_{\text{ox}} \delta_{\text{red}}}{D_{\text{red}} \delta^{\text{ox}}} + \frac{RT}{zF} \ln \frac{j_{d,a} - j}{j - j_{d,c}} \qquad (2.50)$$

Very often only one of the components of the oxidation–reduction system (for example, the reduced form) is originally present in the solution, so that $j_{d,c} = 0$ and (2.50) reduces to

$$E = E^0_{\text{ox,red}} - \frac{RT}{zF} \ln \frac{D_{\text{ox}} \delta_{\text{red}}}{D_{\text{red}} \delta_{\text{ox}}} + \frac{RT}{zF} \ln \frac{j_{d,a} - j}{j} \qquad (2.51)$$

In view of the fact that the term $(D_{\text{ox}} \delta_{\text{red}})/(D_{\text{red}} \delta_{\text{ox}})$ is usually close to unity, the half-wave potential of the reversible wave described by equation (2.51) depends primarily on the thermodynamic term $E^0_{\text{ox,red}}$.

Sometimes the limiting current is not of interest, but rather the ascending exponential-shaped part of the voltammogram is considered. This is particularly true at high concentrations of electroactive substances. The polarization curve then obeys the *Tafel equation* for low j_0, i.e.

$$a + b \log |j| \qquad (2.52)$$

Coefficients a and b have a simple significance when, for example, for a cathodic process equation (2.52) is compared with equation (2.39)

$$a = -2.3\frac{RT}{\alpha F}\log j_0$$

$$b = -2.3\frac{RT}{\alpha F}$$

In its empirical form, equation (2.52) was deduced by Tafel from experiments on hydronium ion reduction (or hydrogen evolution) as early as in 1905.

So far the phenomenological description of electrode processes has been considered. Now the mechanistic factors which control the rates of electrode reactions will be discussed. These rates are far less rapid than could be expected for electron tunelling between the electrode and the reacting species alone. About thirty years ago, Professor Heyrovský, who was not reconciled to the idea of a 'slow' electrode reaction, used to say: 'Nonetheless, electrons move with the velocity of light!' He considered electron transfer very fast and tried to explain the slowness of electrode reactions by sluggish follow-up chemical reactions. In fact the limited rate of the electrode reaction does not originate from the velocity of electron motion but from the readiness of the reacting species to accept or donate an electron.

According to the Franck–Condon principle, electron transfer by means of a tunelling mechanism (cf. p. 60) is possible only when the position of the atomic nuclei and chemical bonds is the same for the reactant particle as for the reaction product. Consider again the example of oxidation of Fe^{2+} and reduction of Fe^{3+} and the equilibrium between the ions and their surroundings. Both the ions are solvated, which can be expressed by the symbols $Fe^{3+}\cdot aq$ and $Fe^{2+}\cdot aq$. The ferric ion carrying a higher positive charge binds its solvation sheath more strongly than the ferrous ion. Thus, the radius of the solvated ferrous ion is about 0.012 nm greater than that of the ferric ion. Consequently, in the reaction path a convenient structure for the reactant particle is formed, enabling the electron to be transferred according to the Franck–Condon principle. This *reorganization of the solvation sheath* is possible because the solvating molecules unceasingly vibrate about the central ion. The difference in the energy of the solvated ion in its equilibrium state and that after reorganization is termed the reorganization energy. This energy term makes a notable contribution to the total energy which must be supplied to the reacting system to overcome the energy barrier hindering the course of the reaction. This energy barrier is called the *activation energy* (see Appendix C). For a cathodic reaction, for example, the rate constant k_{red} (see equation (2.2)) is given by the relationship

$$k_{red} = k'\exp\left[-E_{red}^{\ddagger}/(RT)\right] \tag{2.53}$$

Here k' is a constant (the 'pre-exponential factor') and $E^{\ddagger}$ is the activation energy. This quantity depends on the terms connected with the final conditions in reacting system, i.e. the standard solvation enthalpy of the reduced form $\Delta H_{s,red}^0$, the ionization potential of the reduced form multiplied by -1, $-I$ (cf. p. 15), on the terms connected with the initial conditions, i.e. the electrochemical potential of the electron in the metal (the energy of the Fermi level, see p. 65), $\tilde{\mu}_e = \mu_e^0 - F\varphi$, converted to the enthalpy term by the entropy term of the emission process $+ T\Delta S_e$ and the solvation enthalpy of the oxidized form $\Delta H_{s,ox}^0$, and on the reorganization energy E_r

$$E_{red}^{\ddagger} \approx \tfrac{1}{2}\Delta H_{s,red} - \tfrac{1}{2}I - \tfrac{1}{2}(\mu_e^0 + T\Delta S_e - F\varphi)$$
$$- \tfrac{1}{2}\Delta H_{s,ox} + E_r/4 \tag{2.54}$$

Obviously the value of the charge transfer coefficient from equation (2.54) is $\tfrac{1}{2}$. This equation will be deduced in an outline in Appendix C.

A further experiment involves electrolysis of the solution described on p. 91 with increasing additions of chloride ions (as NaCl or HCl). The rate of both Fe^{2+} oxidation and Fe^{3+} reduction increases in the presence of even very low concentrations of chloride ions (Fig. 51) as a result of adsorption of chloride ions on the platinum surface.

Adsorption is the accumulation of a species on the surface between two media or phases (an interface). Adsorption at the metal/gaseous phase or metal/liquid phase interface is characterized according to the type of bond between the metal and the adsorbed substance. Rather weak physical adsorption is due to van der Waals forces. Then the particles are bound to the metal surface as well as released from it (desorbed) at a high rate. In chemisorption a chemical bond is formed between the metal surface and the adsorbed particle (the adsorbate). This bond is formed and broken rather slowly. The amount adsorbed at the interface increases with the concentration of the substance in the adjacent medium as long as the interface is not fully saturated by the adsorbate. The dependence of the amount adsorbed per unit area of the interface on the concentration is termed the adsorption isotherm.

The electrons not only 'jump' from the metal to ferric ions but are transferred through the adsorbed chloride ions (the 'electron bridges'). In this way the electron transfer is made easier or, in other words, the electrode reaction is

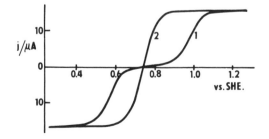

Fig. 51. Voltammograms of 2.7×10^{-3} M Fe^{3+} and 2.4×10^{-3} M Fe^{2+} in 0.5 M $HClO_4$. Rotating gold electrode, rotation speed 2300 min^{-1}, chloride concentration: 1, 0; 2, 10^{-5} mol.dm^{-3}. According to J. Weber, Z. Samec, and V. Mareček

catalysed. The acceleration of an electrode reaction as a result of adsorption of catalysts or of reactants, reaction intermediates, or reaction products is termed *electrocatalysis*. Since various metals adsorb the participants of an electrocatalytic reaction with different strengths, the catalytic effect strongly depends on the electrode material. Various sorts of electrode pretreatment (e.g. by anodic oxidation under surface oxide formation, by cathodic reduction of an oxide layer, etc. in the simplest cases) strongly change the catalytic properties of the surface.

Recently attempts have been made to fix various reactive groups on the electrode surface in order to control electrode processes in a designed way (a similar approach attained some success in heterogeneous catalysis). An example of such a *chemically* (or *surface*) *modified electrode*[2] is the graphite electrode adjusted for oxidation of the basic member of the biochemical electron transfer chain (see p. 176), dihydronicotinamide adenosine dinucleotide (NADH) to NAD$^+$.[3] Normally, this oxidation does not occur at a graphite electrode. In the modification treatment carboxylic groups are formed at the graphite surface by atmospheric oxidation. The 3-hydroxytyramine molecule

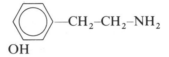

is attached through amidic bonds to the carboxylic groups by the effect of dicyclohexylcarbodiimide and is converted by anodic oxidation to the quinone form. This form readily oxidizes NADH to NAD$^+$ and its resultant reduced form is reoxidized at the electrode. Thus, the 3-hydroxytyramine group acts as a surface electron carrier. A disadvantage of chemically modified electrodes resulting from a decrease in the activity on protracted use strongly limits their applications.

We now give several examples of typical electrode reactions.

Electrolytic *hydrogen evolution*[4]

$$2 H_3O^+ + 2e = H_2 + 2 H_2O \tag{2.55}$$

was for a number of decades the electrode reaction most studied in theoretical electrochemistry.* The hydrogen overpotential was first defined as the value of the voltage imposed on electrodes when hydrogen bubbles just started to evolve. In 1905 J. Tafel formulated the quantitative relationship between the overpotential of hydrogen ion reduction and the current density (2.52). In the subsequent systematic research it was shown that the overpotential strongly depends on the cathode material and the actual course of the hydrogen evolution is more complicated than might follow from equation (2.55) because it is a typical electrocatalytic process. Gaseous hydrogen alone reacts with various metals such that its molecules dissociate to atoms which are subsequently bound to the metal

---

* This was the period when electrode processes were considered at the end of textbooks or chapters on electrochemistry and where no quantitative aspects but rather only vague hypotheses on the course of electrode processes were presented.

by a *chemisorption bond*. The adsorption of gaseous hydrogen occurs according to the equation

$$H_2 \rightleftarrows 2\,H_{ads}$$

i.e.

$$H_2 + \text{surface} \rightleftarrows 2(H \ldots \text{surface}).$$

When changing the type of metal the equilibrium between gaseous hydrogen molecules and adsorbed hydrogen atoms is shifted to different degrees to the side of adsorbed hydrogen, in other words, different metals bind hydrogen with different strengths. Negligible hydrogen adsorption is characteristic of mercury, lead, thallium, and cadmium. Medium adsorption effect is observed for the platinum group metals. Very strong adsorption is exhibited, for example, by tungsten, whose surface is almost completely covered by hydrogen atoms even if the hydrogen pressure is very low.

It is not unexpected that the adsorption of hydrogen plays a major role in electrolytic reduction of hydronium ions. In the first step in this process adsorbed hydrogen atoms are formed by the *Volmer reaction*

$$H_3O^+ + e = H_{ads} + H_2O \tag{2.56}$$

The next step is connected with the removal of adsorbed hydrogen from the electrode surface. This can proceed via the Tafel reaction

$$2\,H_{ads} = H_2 \tag{2.57}$$

or via the Heyrovský reaction

$$H_{ads} + H_3O^+ + e = H_2 + H_2O \tag{2.58}$$

The strength of the adsorption bond decides (i) the slowest step which determines the overall rate of hydrogen evolution, and (ii) the reaction path (reaction mechanism) of hydrogen evolution.

As for mercury, the adsorption bond is weak, reaction (2.56) takes place slowly and requires a large additional amount of electrical energy in order to occur, i.e. the reaction needs a high overpotential. This reaction is then the slowest step in the overall electrode process while it is not easy to distinguish between reactions (2.57) and (2.58) as a follow-up reaction, the latter being more frequently accepted as the step actually occurring. With platinum, reaction (2.56) takes place rather rapidly and has equilibrium character. Because of the medium strength of the adsorption bond, recombination of hydrogen atoms (2.57) occurs relatively easily. This process is then the rate-determining step in hydrogen evolution. Tungsten binds hydrogen atoms very strongly so that breaking two adsorption bonds in order to form a hydrogen molecule would require excessively high energy. Therefore, reaction (2.58) occurs preferentially as only one bond, H–W, need be broken.

Electrolytic hydrogen evolution is important for the production of very pure hydrogen. It is also supposed to be the basic step in the 'hydrogen economy'.[5] In this, still rather hypothetical, project for utilization of nuclear energy, the heat gained from the nuclear pile is transformed into electrical energy in the conventional way. The electrical energy would not then be fed into the electrical grid but would be used for hydrogen production. Instead of electrical current, hydrogen distributed by pipelines like natural gas would then become the main energy carrier. Incidentally, the cost of such a pipeline system is comparable to the cost of the high-voltage transmission system. The local conversion of hydrogen energy to electricity would occur in gas-turbines, in fuel cells (see p. 87), or by direct burning. Personally, I do not think that the 'hydrogen economy' would become a general energy system in the future while, on the other hand, pure hydrogen, the production and storage of which has achieved considerable improvement, particularly in connection with the space programme in recent years, will be rather widely used in future technologies as a powerful and clean reductant.

In the production of heavy water, D_2O, electrolytic hydrogen evolution also plays an important role. The gaseous hydrogen formed by electrolysis contains a higher proportion of light hydrogen than the original water so that the solution is enriched with respect to heavy hydrogen. This enrichment is advantageous as a preliminary step in the overall separation of heavy water.

In a similar way as in the hydrogen evolution reaction, oxygen reduction also consists of several reaction steps.[6] On some electrode materials oxygen reduced to hydrogen peroxide, which is then reduced to water at more negative potentials

$$O_2 + 2H^+ + 2e = H_2O_2 \qquad (2.59)$$
$$H_2O_2 + 2H^+ + 2e = 2H_2O \qquad (2.60)$$

Oxygen reduction occurs at a mercury electrode via this two-step process. In fact, oxygen is reduced to hydrogen peroxide here in a series of partial steps

$$\begin{aligned} O_2 + e &= O_2^- \\ O_2^- + H^+ &= HO_2 \\ HO_2 + e &= HO_2^- \\ HO_2^- + H^+ &= H_2O_2 \end{aligned} \qquad (2.61)$$

but no further complications occur in the absence of a catalyst. A somewhat different situation is encountered with a silver electrode where the presence of a silver oxide on the electrode surface has a strong impact on the reaction mechanism. Hydrogen peroxide reduction then occurs at more positive potentials than reduction of oxygen to hydrogen peroxide. As a result, oxygen is reduced directly to water

$$O_2 + 4H^+ + 4e = 2H_2O$$

In the absence of the oxide, the reduction of hydrogen peroxide takes place in the same way as at a mercury electrode. When hydrogen peroxide is present alone in an alkaline solution, it is simultaneously oxidized to oxygen (reversed equation (2.61)) and reduced to water (equation (2.60)) so that a sort of electrochemical disproportionation* takes place

$$2\,H_2O_2 \rightarrow O_2 + 2\,H_2O$$

When ferrous ions are present in the solution during oxygen reduction they react with the hydrogen peroxide formed at the electrode during its diffusion from the electrode

$$Fe^{2+} + H_2O_2 + H^+ = Fe^{3+} + H_2O + OH^{\cdot}$$

The ferric ions produced in this way are again reduced at the electrode and, in this cyclic process, catalyse oxygen reduction to water. Sometimes, in a more complicated way, but usually with a much more profound effect, catalytic electrochemical reductions of oxygen occur in the presence of metal complexes,[8] particularly of haemoproteins* like haemoglobin or catalase or their prosthetic group hemin.

The most important industrial electrochemical processes are aluminium production, electrochemical plating (particularly of automobile parts), and brine electrolysis where chlorine and sodium hydroxide are formed. All three industrial processes are based totally or partially on *electrochemical metal deposition*.

From a mechanistic point of view the simplest case of deposition of metals occurs at a mercury electrode under amalgam formation. The dependence of the current density on the electrode potential is described by equation (2.43). In this way the cathodic process in the electrolysis of NaCl occurs at a mercury cathode (chlorine is evolved on the graphite or platinized titanium anode). The competing cathodic reaction of hydrogen evolution (here by reduction of water molecules) occurs to only a rather small degree as the hydrogen overpotential is very high. The sodium amalgam formed in this way is then decomposed by the mechanism described on p. 108.

In a more complicated manner metals are deposited on other solid electrode materials (mercury on platinum or platinum on graphite, for example) or at electrodes made of the same solid metal. These processes are governed by the laws of formation and growth of a new phase. When solid metals are deposited the whole process is termed *electrocrystallization*.[9]

The basic rule for solid metal deposition (as for any crystallization) requires the particles to be deposited most easily in steps on the crystal surface where one layer of the crystal lattice covers the underlying layer and, particularly, at kinks formed in these steps (see Fig. 52). For a newly deposited metal atom the energetically most favourable position is that with the greatest contact with the metallic phase already present. This is achieved at a kink in the step (called the half-crystal position). In particular, crystal growth is made easier by the presence of screw dislocations in the crystal lattice. Then the step on the crystal surface does not have rectangular form as the higher atomic layer is slightly sloping so that it

---

* In a disproportionation a reactant is transformed to two products, one of which is in a higher and the other one in a lower oxidation state.
* Haemoproteins are enzymes containing a porphyrin complex of iron as their active site (a prosthetic group) (see p.79).

Fig. 52. A model of a 'half-crystal' position. The crystal surface is only partly covered with deposited metal atoms. The last row of deposited atoms is incomplete. The first vacant position in this row is called the half-crystal position. During crystal growth the first atom settles in this position because here it has the lowest energy, being surrounded on three sides by other atoms. By courtesy of E. Budevski

enters the next lower layer (see Fig. 53). This strange sort of step remains on the surface of a crystal during crystallization so that it rotates around the point where it disappears in the layer above. The crystallization then occurs in a characteristic spiral manner (see Fig. 54). As a result, a low-sloped pyramid of the deposit is formed (Fig. 55).

At sites where more numerous steps and dislocations were present on the original metal surface, the metal is deposited very intensely. Coarse grains (crystallites) are formed and, sometimes, the deposit grows in shrub-shaped aggregates (dendrites, Greek *dendron* = tree—see Fig. 56). In industrial metal deposition these phenomena are rather unwelcome, particularly in electrolytic plating. Therefore, rapid local growth of the deposited metal can be inhibited by various additives which adsorb on active sites of the electrode. The resulting deposit then has a microcrystallinic, consistent structure and its surface attains a bright lustre. Sometimes the growth of a small crystal in only one direction can be achieved using additives. Then a very long and very hard crystal, a 'whisker', is formed.

Anodic oxidation of solid metals follows similar laws as electrocrystallization. It occurs most easily at steps and dislocations in the metal surface. The products are either metal ions dissolved in the electrolyte or insoluble species, particularly oxides, which remain on the surface as films. These films (mainly when they are porous) often make further oxidation possible. Sometimes, however, particularly with chromium, nickel, tantalum, titanium, aluminium and, under certain

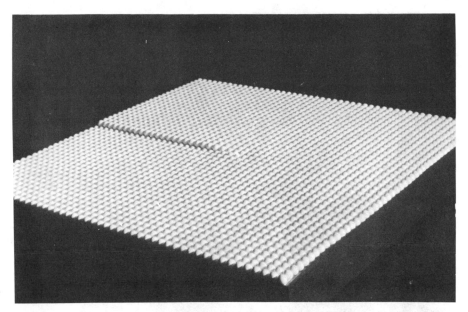

Fig. 53. Crystal growth in a screw dislocation. The crystal face is slightly sloping so that a new crystal layer enters the lower layer and a wedge-shaped step is formed. For crystal growth it is not necessary to start in a half-crystal position but in the initial point of the step. By courtesy of E. Budevski

Fig. 54. During further crystal growth a new wedge is formed in a direction perpencicular to the original step. This rotation of the step results in spiral growth of the crystal. The steps on the spiral are very low so that this kind of growth results in pyramids with very low slopes. By courtesy of E. Budevski

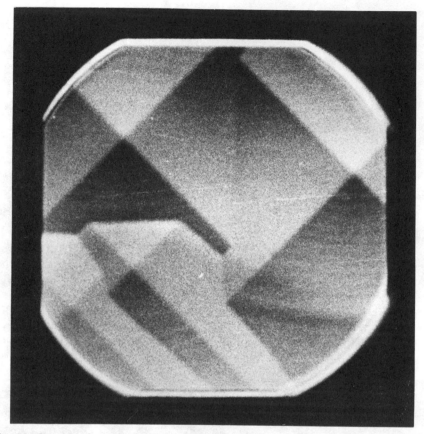

Fig. 55. Growth of pyramids of deposited silver with a very small steepness obtained during electrocrystallization of silver from a $AgNO_3$ solution. Microphotograph by E. Budevski (magnification 700 ×)

conditions, iron, they form a compact layer and bring the metal into a *passive state*. The resulting oxide film is quite often a solid electrolyte with a very low conductivity and protects the metal from further oxidation.

An interesting application of anodic metal dissolution is *electrochemical machining*,[10] discovered by Gusseff in 1928 but introduced into technology after 1960. Here, the cathode of the cutting tool is kept at a minimum possible distance from the machined material (which is an anode) so that, after switching on the electric current, a mirror image of the tool is formed in the material by anodic dissolution. The electrolyte solution flows rapidly through the narrow space between the anode and the cathode to decrease the overheating of the electrode and to remove the products of the anodic process. Both parts are kept at a constant distance in order to compensate the loss of the anode. In this way arbitrary profiles can be formed in the machined material. The second advantage, in comparison with traditional machining, is that a great variety of metallic

Fig. 56. Dendrite growth of copper from a $CuSO_4$ solution in dilute H_2SO_4. Microphotograph by A. R. Despić, G. Savić-Maglić, and M. Jačović (magnification 5000 ×)

materials can be machined since this method does not depend, for example, on the metal hardness. A scheme of electrochemical drilling is shown in Fig. 57.

Metal deposition from its coordination compounds (complexes) requires some additional electrical energy which corresponds at least to the energy of formation of the complex of its components, the free metal ions and the coordinated particles (ligands). The participants in the electrode reaction are often the particles of the complex while sometimes the complex first has to dissociate to its

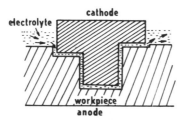

Fig. 57. Electrochemical machining

components. This reaction may not be fast enough and its course then determines the current value. (See J. Koryta, in ref. 8 below.)

Consider once again the sodium amalgam formed by electrolysis of NaCl (p. 103). In contact with water sodium amalgam behaves similarly to, but less violently than, metallic sodium. This would appear to be simply a chemical reaction

$$2\,NaHg_x + 2\,H_2O = H_2 + 2\,NaOH + 2xHg \tag{2.62}$$

However, the reaction of sodium amalgam with water is not this simple. When the current is interrupted during the electrolysis of NaCl (or when sodium amalgam comes into contact with a solution of NaCl) the electrode acquires an instantaneous equilibrium potential $E_{Na^+/NaHg_x}$. At this electrode potential water is reduced

$$2\,H_2O + 2e = H_2 + 2\,OH^-$$

In order to preserve electroneutrality this cathodic process must be compensated by an anodic reaction which obviously is the ionization of sodium

$$Na = Na^+ + e$$

The compensation of the processes can be clearly seen from their polarization curves (see Fig. 58). The electrode potential where the anodic and the cathodic currents cancel each other is termed a *mixed potential* because it is determined by two completely different electrode processes, in contrast to an equilibrium

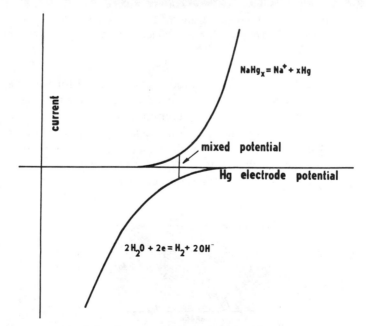

Fig. 58. Decomposition of sodium amalgam and the establishment of a mixed potential

electrode potential where a single reversible electrode reaction occurs. The dissolution rate is indicated by the anodic current flowing at the mixed potential which is called the *corrosion current*, analogous to the exchange current (cf. p. **00**) flowing at the equilibrium electrode potential. The oxidations of metals in solution or in a moist environment, termed *metal corrosion*,[11] occurs by the same mechanism. The cathodic process is usually hydronium ion or oxygen reduction. The mechanism of metal corrosion in a gaseous environment is sometimes different since it is based on the direct reaction of the metal with oxygen.

The corrosion of metals leads to enormous economic losses. It considerably limits the lifetime of metallic constructions, pipelines, boilers, automobiles, etc. It is often impossible to reduce corrosion by modifying the medium where the metallic parts are placed. Therefore protection against corrosion is achieved by a suitable composition or surface treatment of the metallic material. The choice of the method of protection is based on the principles of electrochemical kinetics (in spite of the fact that it was usually carried out quite empirically in the past). Anti-corrosive steel is obtained by alloying with nickel or chromium. Both these metals easily form anodic films converting the material into a passive state. Nickel or chromium can be deposited on a steel surface with an analogous result. Protective films can be obtained by treating the surface with phosphoric acid. In addition to slowing down the dissolution rate, these films prevent the electroactive species in the cathodic reaction (hydronium ions, oxygen) from reaching the surface by diffusion or inhibit their electrode reactions. The coating of the surface by organic materials, particularly of polymeric character, is carried out for similar reasons.

References

1. For general information on electrode kinetics see P. Delahay, *Double Layer and Electrode Kinetics*, John Wiley & Sons, New York, 1965.
   B. E. Conway, *Theory and Principles of Electrode Processes*, Ronald Press, New York, 1965.
   For tables of electrode reaction parameters see N. Tanaka and R. Tamamushi, *Electrochim. Acta*, 9 (1964) 963.
   K. J. Vetter, *Electrochemical Kinetics*, Academic Press, New York, 1967.
2. P. R. Moses, L. Wier, and R. W. Murray, *Anal. Chem.*, **47** (1975) 1882.
   K. D. Snell and A. G. Keenan, *Chem. Soc. Rev.*, **8** (1979) 259.
3. D. C. Tse and T. Kuwana, *Anal. Chem.*, **50** (1978) 1315.
4. A. N. Frumkin, 'Hydrogen overvoltage and adsorption phenomena. Part I', in: *Advances in Electrochemistry and Electrochemical Engineering*, Vol. 1 (P. Delahay and C. W. Tobias, eds.), Wiley–Interscience, New York, 1961, p. 65; Part II, Vol. 3, 1963, p. 267.
   L. I. Krishtalik, 'Hydrogen overvoltage and adsorption phenomena. Part III', in: *Advances in Electrochemistry and Electrochemical Engineering*, Vol. 7 (P. Delahay and C. W. Tobias, eds.), Wiley–Interscience, New York, 1970.
5. D. P. Gregory, *Scientific American*, **228**, No. 1 (1973) 13.
6. J. Hoare, *The Electrochemistry of Oxygen*, Wiley–Interscience, New York, 1968.
7. M. Březina J. Koryta, and M. Musilová, *Coll. Czech. Chem. Comm.*, 33 (1968) 3397.
8. J. Heyrovský and J. Kůta, *Principles of Polarography*, Academic Press, New York, 1966.

J. Koryta, 'Electrochemical kinetics of metal complexes', in: *Advances in Electrochemistry and Electrochemical Engineering*, Vol. 6 (P. Delahay and C. W. Tobias, eds.), Wiley–Interscience, New York, 1967, p. 289.

9. M. Fleischmann and H. R. Thirsk, 'Electrocrystallization and metal deposition', in: *Advances in Electrochemistry and Electrochemical Engineering*, Vol. 2 (P. Delahay and C. W. Tobias, eds), Wiley–Interscience, New York, 1963, p. 123.

10. J. P. Hoare and M. A. LaBoda, 'Electrochemical machining', in: *Comprehensive Treatise in Electrochemistry*, Vol. 2 (J. O'M. Bockris, B. E. Conway, E. Yeager and R. White, eds.), Plenum Press, New York, 1980.

11. H. H. Uhlig, *Corrosion and Corrosion Control: An Introduction to Corrosion Science and Engineering*, 2nd edn, John Wiley & Sons, New York, 1971.

Organic Electrochemistry

Organic compounds are much more numerous than inorganic species, those already known exceeding two million in number. In the same proportion is the number of the electrode reactions of organic substances to that of the inorganic processes.[1] On cathodes of various materials are reduced double bonds between carbon atoms, in particular conjugated double bonds, the nitro, halogen and carbonyl groups, pyridine and other heterocyclic nuclei, etc. At anodes, not only simple oxidations but also chlorinations and cyanations take place and even carboxylic acids can be converted to hydrocarbons and carbon dioxide in the Kolbe reaction. (In fact, first described in M. Faraday.)

Electrosynthetic methods have some advantages, for example the yield of an electrochemical reaction taking place at constant potential is often large, but they are restricted to laboratory work. In industrial production they have two main drawbacks: they proceed rather slowly since they are confined to the electrode/electrolyte solution phase boundary and they consume expensive electric energy. The only large-scale electrochemical process in the field of organic chemistry is reductive dimerization of acrylonitrile to adiponitrile in the sequence of reactions

$$CH_2{=}CH{\cdot}CN + 2e = \bar{C}H_2{-}\bar{C}H{\cdot}CN$$
$$\text{acrylonitrile}$$

$$\bar{C}H_2{-}\bar{C}H{\cdot}CN + CH_2{=}CH{\cdot}CN$$
$$= NC{\cdot}\bar{C}H{\cdot}CH_2{\cdot}CH_2{\cdot}\bar{C}H{\cdot}CN$$
$$NC{\cdot}\bar{C}H{\cdot}CH_2{\cdot}CH_2{\cdot}\bar{C}H{\cdot}CN + 2H_2O$$
$$= NC{\cdot}(CH_2)_4{\cdot}CN + 2OH^-$$
$$\text{adiponitrile}$$

Adiponitrile is the basic material in the synthesis of Nylon 606. (In a dimerization two molecules of the reactant, a monomer, are converted into one molecule, a dimer.)

In the organic chemistry laboratory various compounds are reduced by metals or metal amalgams. These reactions are considered as 'purely' chemical, in the same way as reaction (2.62). In fact they occur via a corrosion mechanism. Thus, for example, when nitrobenzene is reduced by zinc powder, its molecules are reduced by electrons from the metal and, at the same time, the zinc ions are formed by anodic oxidation of zinc.

References

1. M. M. Baizer, Ed., *Organic Electrochemistry*, Marcel Dekker, New York, 1973.
2. R. A. Adams, *Electrochemistry at Solid Electrodes*, Marcel Dekker, New York, 1969.

Electric Double Layer

As already mentioned on p. 64, the electrode is not only the site for charge transfer from the electrode to the solution and back but also is a sort of a molecular condenser. Now the properties of this *electric double-layer*[1] will be considered.

An alkaline solution containing 0.1 M sodium sulphite and 0.1 M NaOH is poured into a glass beaker with a layer of mercury at the bottom. The solution is connected to a saturated calomel electrode by means of a salt bridge. Pure argon or nitrogen bubbles through the solution while the mercury electrode is polarized by a voltage of -0.7 V as a cathode; the saturated calomel electrode is the anode. In this way a small amount of mercuric ions formed in the solution during contact with mercury in the presence of traces of atmospheric oxygen, which was not reduced by the sulphite, is reduced to metallic mercury. After some time the electric current falls to zero. The solution then contains only sulphite, hydroxide, and sodium ions together with a small concentration of sulphate ions formed by oxidation of sulphite by oxygen. Under these conditions the system remains in a currentless state. When the voltage is increased to -1.0 V a large instantaneous current flows through the circuit and immediately falls to zero. After that a potential of -1.0 V is maintained with no further supply of electric charge. The initial current obviously is the *charging current* (see p. 64). The negative potential can be increased to a value of -1.5 V with the same result—on charging from an external source the electrode acquires a definite potential value which is maintained for an arbitrary period of time. Thus, it behaves like a condenser without leakage. It is called an *ideally polarized electrode*.

A similar experiment could be carried out with a KCl or KF solution but the difficulties with removing oxygen from the solution prevent it being carried out without tedious precautions. Oxygen is reduced at the electrode (naturally, also in the situation when the external voltage source is disconnected) and discharges the molecular condenser (it decreases the negative charge of the metal by removing the electrons). When no current is supplied from the external source the negative potential of the electrode decreases. However, some of the properties of the electric double-layer can be successfully investigated, even in the presence of trace concentrations of oxygen.

When, for example, 0.1 M KF solution completely freed of oxygen and other impurities is poured over a mercury electrode, the electrode acquires a potential of -0.19 V versus the standard hydrogen electrode. Before coming into contact with the electrolyte, the electrode possesses a negligibly small charge because the capacity of a metal in contact with a gas or vacuum is extremely low. In contact with the electrolyte (with no species present which would react at the electrode at -0.19 V), the electrode can accept no charge from it. The potential of -0.19 V versus the standard hydrogen electrode is the *zero charge potential*.

The zero charge potential strongly depends on the electrolyte composition which is connected with the ability of individual types of ions to become adsorbed at the electrode/electrolyte solution interface. Fluoride, potassium, or sodium ions are not adsorbed at all. When the tendency towards adsorption is greater for the anion of the electrolyte than for the cation, the zero charge potential is more negative than in KF or NaF solutions, e.g. in solutions of alkali metal chlorides, bromides, or iodides. On the other hand, when the adsorption of the cation is stronger than the anion (in a solution of tetrabutylammonium chloride, for example) the zero charge potential is more positive. The zero charge potential depends markedly on the electrode material (for silver, for example, it is about 0.5 V more positive than for mercury).

In order to charge the electrode to a more positive potential, a supply of positive charge is required and, similarly, a supply of negative charge for negative potentials. The amount of electricity necessary for charging the electrode depends on its capacity which, at potentials more negative than the potential of zero charge, is about $20 \mu\text{F.cm}^{-2}$ in simple inorganic electrolyte solutions, while at more positive potentials it is higher. The electrode capacity is not constant but depends more or less on the potential of the electrode. In order to include this dependence the differential capacity of the electrode, $C = \text{d}q/\text{d}E$ is introduced (q is the charge of the electrode). These values are very high because of the extremely small thickness of the molecular condenser (cf. equation (1.2)). The simplest picture of the electric double layer is that of a plate condenser where one armature is represented by the charges in the metal and the other one being formed by the ions at a minimum distance in the solution (this distance is really very small, of the order of tenths of a nanometer). In this structure the electroneutrality rule is not preserved and a definite type of ion predominates in the electric double-layer region of the solution. As a whole, however, the electric double-layer is electroneutral, the sum of the charge in the metallic and in the solution part of the double layer being equal to zero.

This picture of the electric double layer (introduced by H. L. F. von Helmholtz in 1879) is, however, valid for a rather concentrated electrolyte solution. The more dilute the electrolyte the larger the distances will be where the ions forming the solution part of the electric double-layer can escape by thermal motion. A structure which is similar to the ionic atmosphere around individual ions (cf. p. 26) then predominates. This is called the *diffuse double-layer*. Fig. 59 shows the structure of the electric double-layer at the metallic electrode/electrolyte solution interface. The thickness of the diffuse double-layer strongly depends on the electrolyte concentration. In electrolytes consisting of a monovalent cation and a monovalent anion it has a thickness of an order of tenths of a nanometer at 0.1 molar concentrations, whereas the thickness becomes as high as tens of a nanometer at millimolar concentrations.

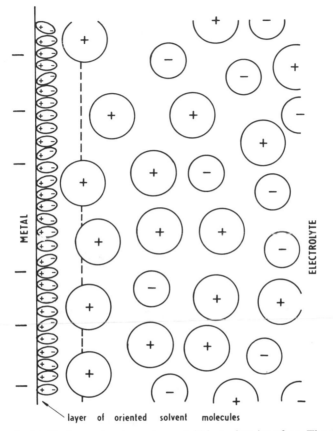

Fig. 59. Electric double-layer structure at a metal/electrolyte interface. The dashed line indicates the plane reached by the ion centres at minimum distance from the electrode (the 'outer Helmholtz plane'). The absence of ion adsorption is assumed since otherwise this structure would be more complicated. The circles correspond to solvated ions (greater solvation of cations than of anions is assumed)

Ideally polarized electrodes can be attained in few electrode systems, particularly with regard to the electrode metal. On the other hand, electrodes on which an equilibrium potential is established have an electric double-layer of the same character. This is an important fact, e.g. for *semiconductor electrodes* which are corroding systems in most media. A change in the electric double-layer properties is then achieved through a change in the composition of the oxidation–reduction system present in the solution. In contrast to metallic electrodes where the charge of the metallic part of the electric double-layer is homogeneously dispersed over the metal surface and the concentration of charge carriers is high, semiconductor electrodes have a diffuse double-layer even in the inside of the semiconductor because of the limited value of its permittivity and low charge-carrier concentration. In the terminology of solid-state physics, the diffuse double-region is termed the space-charge region (Fig. 60).

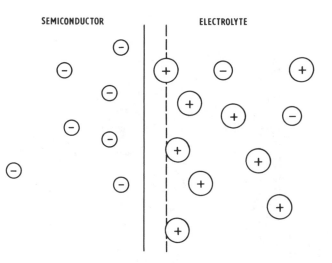

Fig. 60. Electric double-layer structure at a semiconductor/electrolyte interface. A diffuse double-layer is formed in the electrode as well as in the solution

The electric double-layer appears even at the electrolyte solution/insulator interface if the electric charges are fixed, e.g. by transfer of ions of a definite charge from the surface of an ionic crystal into the solution, by ion adsorption, by the presence of ionizable groups in polymeric materials, etc. The formation of the electric double-layer has a great impact on the transport processes with participation of charged colloidal particles, of porous and of other membrane systems (see p. 35) all of which are *electrokinetic phenomena.*

According to their charge, colloidal particles in an electrolyte solution move in the direction of the electric field imposed on the system or against it. This phenomenon is termed *electrophoresis.* Their velocity increases when the concentration of the indifferent electrolyte decreases. This effect shows that electrophoresis is connected with the presence of the diffuse double-layer at the surface of the particles. Under the influence of the electric field the space charge in the diffuse double-layer is set in motion in the opposite direction to the movement of the particle itself. If the charge is spread to large distances from the interface, which is the case in dilute solutions, the particle exerts only a small hydrodynamic resistance to the movement of the charge in the diffuse double-layer. On the other hand, at high electrolyte concentrations, the electric double-layer is reminiscent of a plate condenser with the solution charge fixed on the plane facing the surface of the solid phase. Under these conditions it is impossible to induce a movement of the solution charge from the interface, and electrophoresis ceases. An inverse process to electrophoresis is observed when charged colloidal particles dispersed in an electrolyte solution sink to the bottom. An electrical potential difference called the *sedimentation potential* is then measured between the upper and lower layers of the solution. During sinking the diffuse double-layer lies behind the colloidal particle resulting in an electric potential difference.

The main application of electrophoresis[2] is the separation of proteins. In an electric field various types of these macromolecules move with different velocities because of their size and charge, so that their distribution in the electrophoretic cell can be analysed optically. In preparative electrophoresis, proteins are accumulated in different compartments of the cell.

References

1. R. Parsons, 'Structure of the electrical double layer and its influence on the rates of electrode reactions', in: *Advances in Electrochemistry and Electrochemical Engineering*, Vol. 1 (P. Delahay and C. W. Tobias, eds.), Wiley–Interscience, New York, 1961, p. 1. P. Delahay, *Double Layer and Electrode Kinetics*, John Wiley & Sons, New York, 1965. J. O'M. Bockris, B. E. Conway and E. Yeager (eds.), *The Double Layer*, Vol. 1 of *Comprehensive Treatise of Electrochemistry*, Plenum Press, New York, 1980.
2. C. J. O. R. Morris and P. Morris, *Separation Methods in Biochemistry*, Pitman Publishing, London, 1976. S. Hjirtén, *Free Zone Electrophoresis*, Almquist and Wiksell, Uppsala, 1967.

Photoelectrochemistry

In recent years great interest has arisen concerning various ways of utilizing solar energy. The incident light can be transformed to heat (with considerable loss of energy), which is then either used directly or accumulated by one of several methods. On the other hand, light may be also converted directly into electrical energy, which may also be achieved through several physical and electrochemical methods. In photoelectrochemical energy conversion a suitable galvanic cell is brought to an excited state by the effect of irradiation. This excitation can occur in two ways:

(1) Light excites one or both electrodes of the cell. This phenomenon is termed the *electrochemical photovoltaic effect* (the main subject of this section).

(2) Light affects a solution component which either directly or, more frequently, after a suitable chemical reaction, changes the potential of the adjacent electrode (the *photogalvanic effect*).

Generally, in the photovoltaic effect an electric potential is formed by the irradiation of semiconductor materials. In solid-state physics these phenomena are mainly investigated with photovoltaic cells either with a p–n junction or with a semiconductor/metal junction (Scholtky-type photovoltaic cells). When the energy of the photons of the light illuminating a semiconductor plate is larger than the energy gap of the semiconductor, the electron–hole pairs form under the surface of the plate. This process cannot be utilized in an isolated semiconductor because the pair would immediately recombine, release thermal energy, and the light energy would be dissipated. Photovoltaic effect that can really be observed are distinguished according to the nature of a foreign phase which is connected to the semiconductor under illumination. In electrochemical photovoltaic effects this is an electrolyte solution containing a redox system. When it is in contact with the semiconductor an electrode reaction occurs in order to equalize the Fermi levels of the electrons in the electrode and in the solution. The charging of

116

the electrode connected with the establishment of the electrode potential results in formation of a space charge region (see p. 113). The induction of the photovoltaic effect requires a redox potential of the electrolyte such that the space charge region is formed by the minority carriers, i.e. the holes in an n-type semiconductor and the electrons in a p-type semiconductor. Then the electric field established under the surface of the n-type semiconductor forces the electrons to migrate into the bulk of the semiconductor while the holes are transported to the surface. In this way *charge separation* occurs after the electron–hole pairs have been formed on illumination, which cannot be achieved in an isolated semiconductor (Fig. 61). The holes thus generated charge the electrode to a more positive potential. The difference between the original equilibrium potential and this new value is called the *photopotential*. This deviation of the electrode potential from equilibrium causes the electrode reactions based on hole transfer to the reductant species in the solution to occur

$$\text{red} + \text{h}^+ \rightarrow \text{ox}$$

This can happen only when the process takes place in a galvanic cell: in addition to the semiconductor electrode, another metallic electrode connected with the semiconductor electrode through an electric energy consumer is present in the solution. The electrons originally transported into the bulk of the semiconductor electrode cross the junction to the metallic electrode where an electrode reaction occurs in the opposite direction

$$\text{ox} + \text{e}^- \rightarrow \text{red}$$

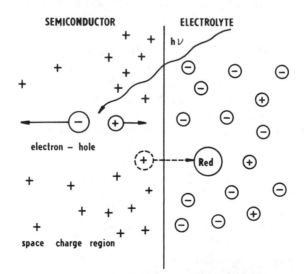

Fig. 61. Formation of an electron–hole pair at an n-type semiconductor/electrolyte interface by incident light. The electric field in the space-charge region (electric diffuse double-layer) drives the electrons into the bulk of the semiconductor and the holes to the surface

In this way an electric current has been generated and the light energy transformed into electric energy. Electrochemical systems containing a semiconductor electrode and a metallic electrode in a solution of a single oxidation–reduction system are termed *electrochemical photovoltaic cells*. They supply electric energy and their composition during service does not change since an identical electrode reaction takes place at each electrode—only in the opposite direction (therefore they are sometimes called regenerative photoelectrochemical cells).

The main disadvantage of these devices is the insufficient stability of the semiconductor electrode resulting from *photocorrosion*. The holes generated at any n-type semiconductor electrode can attack the electrode material instead of reacting with the reductant in the solution. For example, the n-CdS electrode is oxidized according to the equation

$$CdS + 2h^+ \rightarrow Cd^{2+} + S$$

However, this process can be suppressed by a proper choice of the redox properties of the electrolyte. The systems S^{2-}/S_n^{2-}, Se^{2-}/Se_2^{2-}, and Te^{2-}/Te_2^{2-} exhibit a performance suitable for this purpose. Materials used for n-type semiconductor electrodes are sulphides, selenides, and tellurides of cadmium and GaAs, InP, and GaP. The efficiency of light energy conversion extends from 1 to 2% with n-CdS in a S^{2-}/S_n^{2-} electrolyte, to 10% with n-CdTe in Te^{2-}/Te_2^{2-} and 9–10% with n-GaAs in Se^{2-}/Se_2^{2-}.

A *photoelectrolytic cell* differs from the electrochemical photovoltaic cell in that a different electrode reaction takes place at each electrode, resulting in suitable electrolysis products in contrast to the preceding case aimed at electric energy generation.

References

1. A. J. Nozik, *Ann. Rev. Phys. Chem.*, **29** (1978) 189.
   A. J. Bard, *Science*, **207** (1980) 139.
2. H. Gerischer, *J. Electroanal. Chem. Interfacial Electrochem.*, **58** (1975) 263.
   Yu. Ya. Gurevich, Yu. V. Pleskov and Z. A. Rotenberg, *Photoelectrochemistry*, Plenum Press, 1980.

Electroanalytical Chemistry

Electrochemical methods of analysis of solutions have experienced periods of boom and decline. At present quite a few analytical procedures based on various sorts of electrochemical indication have survived. They may be classified by means of several criteria. When a reagent is added to the analysed solution (the sample) and an electrochemical signal (electrode potential, electric current, conductance, etc.) indicates the completion of its reaction with the substance to be determined (a determinand), this analytical approach is called a *titration* method (for example, a conductometric titration, see p. 31). Usually, the volume of reagent solution consumed in the reaction is then measured. Another group of methods is based on complete separation of the determinand by electrolysis of the analysed solution. When the amount (usually of a metal) deposited at an electrode is measured, usually by weighing, the method is termed *electrogravimetry*. *Coulometry* is based on determination of the total amount of electricity consumed for complete electrochemical transformation (reduction with or without deposition, oxidation) of the determinand. Finally, there is a group of methods which interfere with the sample only negligibly or not at all. The determinand concentration is then assessed either on the basis of a single signal or of the dependence between electrical quantities and, in some methods, of the dependence of an electric quantity on time, recorded during the electrochemical measurement. *Potentiometry* works in a currentless state making use of an equilibrium potential or of a mixed potential. Direct potentiometry is at present restricted to membrane systems termed *ion-selective electrodes* (see p. 137). A number of electroanalytical methods are based on *electrolysis with polarizable microelectrodes*. Those electrochemical methods that are of practical value at present will now be considered in greater detail.

The *coulometric method* requires only one reaction to occur at the electrode. For example, metallic silver is deposited on a silver electrode by electrochemical reduction of the silver ions present in the solution. According to the Faraday Law (1.30) for $z_A = 0$ the electric current is directly proportional to the deposition rate of silver (i.e. to the flux of silver ions at the surface of the electrode) and, consequently, the charge consumed for silver ion reduction during a given time is proportional to the amount of silver deposited. If this amount is expressed in mols, the proportionality factor is the Faraday constant. When all the silver present in the solution has been deposited this amount can be determined according to the charge consumed for the deposition using a suitable current integrator. The coulometry at constant electrode potential is one of the most exact analytical methods.

The basic characteristics of *voltammetry* have been already pointed out on p. 88. As indicator electrodes in analytical voltammetry, the dropping mercury electrode (Fig. 62), the hanging mercury drop electrode, and various types of rotating solid metal electrodes, particularly the rotating disc electrode (Fig. 63), are used.

Voltammetry with the dropping mercury electrode was termed '*polarography*' by J. Heyrovský who discovered this method in 1922. The continuous renewal of the mercury surface prevents the surfactants present in the electrolyte solution from adsorbing at the electrode, thus rendering the voltammograms reproducible. The high overpotential of hydrogen evolution at mercury extends the potential 'window', where waves of electroactive substances can be found, to quite negative potentials. In analytical polarography, generally average currents (the amount of electricity consumed during the life of a drop divided by the drop-time) are recorded as a function of the imposed voltage. The average limiting diffusion current at the dropping mercury electrode is governed by the Ilkovič equation

$$\langle I_d \rangle = 0.63 \times 10^{-3} zFm^{2/3} t_1^{1/6} D^{1/2} c_0$$

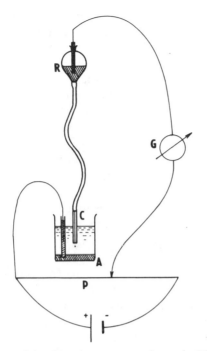

Fig. 62. The original form of the dropping mercury electrode. Mercury preserved in the reservoir R drops from a capillary C into the electrolyte. Mercury on the bottom of the cell functions at the same time as a reference and as an auxiliary electrode. The potential difference is imposed on the electrodes from potentiometer P. According to J. Heyrovský

where z is the charge number of the electrode reaction, F the Faraday constant, m the flow-rate of mercury (g.s^{-1}), t_1 the drop-time (usually several seconds), D the diffusion coefficient (cm^2.s^{-1}), and c_0 the concentration of the electroactive substance. A stationary version of the dropping mercury electrode is the hanging drop electrode where a mercury drop is squeezed out, using a micrometer screw, from a glass capillary attached to a mercury-filled glass syringe.[1]

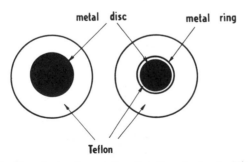

Fig. 63. The rotating disc electrode and the ring-disc electrode. View from below. The electrodes are fixed in a cyclindrical Teflon body

120

Polarography is a very sensitive method but the determination of substances present in micromolar and lower concentration is interfered with by the charging current which is rather large because of the continuous increase in the surface of the dropping electrode. These concentrations (of metal ions, for example) are often of interest because they endanger the human environment. In order to increase the sensitivity of polarographic methods, various techniques have been proposed, the most successful being *pulse polarography* which removes the interference of the charging current by measuring the current of the drop at the end of the drop-life (Fig. 64).[2]

Among solid metal microelectrodes the rotating disc electrode introduced by A. N. Frumkin and V. G. Levich[3] is a favourite tool in electrochemical investigation. A significant property of this electrode is the constant current density at any point of the electrode. The limiting current at the rotating disc electrode is given by the Levich equation

$$I_d = 0.62.10^{-3} zFAD^{2/3} v^{-1/6} \omega^{1/2}$$

where A is the area of the electrode (cm^2), D the diffusion coefficient of the electroactive substance, v the kinematic viscosity (the product of the viscosity coefficient and the density of the solution), and ω the angular velocity of the electrode.

Enormous sensitivity is achieved by *anodic stripping analysis*[4] where a trace metal present in the sample is deposited for a definite, rather long time on a rotating disc electrode or hanging mercury drop electrode. Then the potential of the electrode is switched to a value where the deposited metal is anodically reoxidized. The amount of electricity consumed in the reoxidation indicates with great precision the concentration of the determinand in the solution.

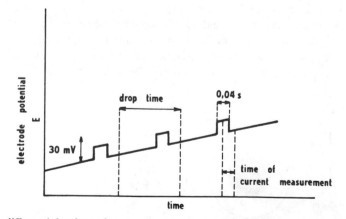

Fig. 64. In differential pulse polarography a basic slowly increasing voltage is imposed on the dropping electrode. Before drop detachment a small voltage pulse is fired on the electrode. The difference of the integrated current flowing during the second half of this pulse and of the integrated current during the same period of time after the pulse is recorded in dependence on the basic voltage

For analysis of a soluble product of an electrode reaction the ring-disc electrode[5] has been designed (Fig. 63). The products of an electrode reaction that occurs at a definite potential on the disc are transported to the ring during rotation of the electrode. The ring potential is different from the disc potential so that, for example, the products of an oxidation on the disc are reduced on the ring. Among other applications, unstable intermediates of electrode reactions, which otherwise are converted to inactive products by chemical reactions taking place in the solution, can be determined in this way.

In some electroanalytical determinations the current is measured at constant potential. Among these *amperometric methods* the most important is the determination of oxygen in blood and tissues. Early attempts to measure the oxygen content with a bare platinum electrode impaled into the organ were completely unsuccessful because the components of the biological fluid, particularly of a high-molecular nature, were adsorbed on the electrode and inhibited oxygen reduction. These obstacles have been removed in the Clark membrane electrode[6] (Fig. 65). Here the platinum indicator microelectrode together with the reference electrode placed in a suitable electrolyte solution is separated from the sample by a hydrophobic porous Teflon film. Oxygen dissolved in the sample diffuses to the electrode compartment through the gas-filled pores of the film and is reduced at the electrode. Other voltammetric methods have already been described on p. 88.

In the methods described so far, a constant or time-dependent potential is imposed on the indicator electrode from an external source. All these methods are termed *potentiostatic*. An electrode with a stationary surface placed in an unstirred solution can also be polarized with a constant current. This is a *galvanostatic method* or *chronopotentiometry*, where the dependence of the

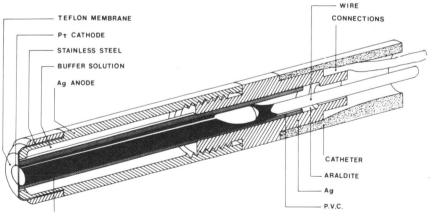

Fig. 65. Catheter oxygen electrode (Clark type) of 1.2 mm external diameter with a ring-shaped platinum cathode about 3 μm thick. This type of electrode is used for the determination of arterial oxygen concentrations. From P. Kreuzer, H. P. Kimmich, and M. Březina, with permission

electrode potential on time is recorded. In this sort of electrolysis the constant current density forms a constant concentration gradient of the electroactive substance at the electrode surface. The substance reacting at the electrode is gradually exhausted in the neighbourhood of the electrode, so that its concentration at the surface of the electrode decreases. When it approaches zero, the original electrode reaction cannot supply a sufficient amount of substance to satisfy the imposed current density and a transition of the electrode potential occurs to reach a value where another electrode reaction takes place. The time until this rapid potential change begins is termed the transition time[7] (Fig. 66).

Most of the methods discussed are suitable for analysis as well as for research on mechanisms and rates of electrode reactions. The accuracy of the coulometric method and the sensitivity of pulse polarography and of anodic stripping analysis (with a detection limit as low as $10^{-9}\,\mathrm{mol.dm^{-3}}$) classifies them among prominent methods of analytical chemistry.

For investigation of electrode reactions there are a number of other modes of electrode polarization. Thus, for example, a low amplitude a.c. current can be applied to an electrode placed in a solution containing both the reductant and oxidant forms of a redox system. The amplitude and the phase shift of the resulting small a.c. voltage characterize the impedance of the electrode, which consists of two components. One of them depends on the rate of the electrode reaction and on the diffusion rate of the components of the redox system to and from the electrode. This component is called the *faradaic impedance*. The other components are identical with the capacitance of the electrode.[8]

In the *coulostatic method* the electrode in equilibrium with a redox system is charged from a small condenser. In this way it is shifted from equilibrium for a short period of time and then it returns to the original potential at a rate depending on the rate constant of the electrode reaction. Since the time-dependence of the electrode potential is measured in a currentless state (the

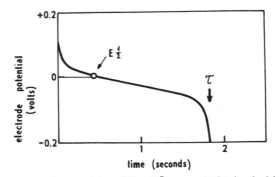

Fig. 66. The chronopotentiogram (plot of $E - E^0$ versus t) obtained with constant current density $j = -10^{-2}\,\mathrm{A\,cm^{-2}}$ and a solution containing an oxidant $(c_{ox} = 5 \times 10^{-2}\,\mathrm{mol.\,dm^{-3}}, D_{ox} = D_{red} = 10^{-5}\,\mathrm{cm^2s^{-1}})$. The transition time is denoted as τ. The electrode reaction is very fast ('reversible', cf. p. 000) so that at $t = \tau/4$ the electrode potential is equal to the half-wave potential $E_{1/2}$

electrode reaction removing the necessary charge from the electrode), this method is not complicated by the ohmic potential drop.[9]

References

1. J. Herovský and J. Kůta, *Principles of Polarography*, Academic Press, New York, 1966.
   Z. Galus: *Fundamentals of Electrochemical Analysis*, Ellis Horwood, Chichester, 1976.
2. G. C. Barker and A. W. Gardner, *Fresenius Z. Anal. Chem.*, **173** (1960) 79.
   E. P. Parry and R. A. Osteryoung, *Anal. Chem.* **37** (1965) 1634.
3. V. G. Levich, *Physico-Chemical Hydrodynamics*, Prentice-Hall, Englewood Cliffs, 1962
4. F. Vydra, K. Stulík, and E. Juláková, *Electrochemical Stripping Analysis*, Ellis Horwood, Chichester, 1976.
5. W. J. Albery and M. E. Hitchman, *Ring-disc Electrodes*, Oxford University Press, 1971.
6. F. Kreuzer, H. P. Kimmich, and M. Březina, 'Polarographic determination of oxygen in biological materials', in: *Medical and Biological Applications of Electrochemical Devices* J. Koryta, ed.), John Wiley & Sons, Chichester, 1980, p. 173.
7. P. Delahay, *New Instrumental Methods of Electrochemistry*, Interscience, New York, 1954.
8. J. E. B. Randles, 'Concentration polarization and the study of electrode reaction kinetics', in: *Progress in Polarography*, Part I (P. Zuman and I. M. Kolthoff, eds.), Interscience, New York, 1962, p. 123.
9. B. E. Conway, *Theory and Principles of Electrode Processes*, Ronald Press, New York, 1965.

Summarizing Chapters 1 and 2

The electrochemistry of electrolyte solutions is a stable component of physicochemical knowledge, indispensable for research as well as for the solution of many practical tasks in technology, analytical chemistry, biochemistry, and physiology. It is, however, not an exciting field where rapid evolution resulting in far-reaching applications could be expected. On the other hand, the electrochemistry of electrode systems has direct utilization in the practice of technology and analytical and clinical chemistry. In the latter field it must, however, compete with other methods. In the past thirty years several examples of rapidly increasing interest could be observed, an interest which, after some time, began to decline but never disappeared completely. Thus, for example, polarography belonged among the five most frequently used methods of analytical chemistry in the years just after 1945. Since then its popularity has markedly decreased, but interest in trace analysis by modified polarographic methods has survived (together with routine polarographic investigations of the redox properties of substances in organic or inorganic laboratories). At the end of the 1950s fuel cells became a favourite research subject in electrochemistry. The search for light power sources suitable for space vehicles played a major role here. At the same time, however, the older hopes for power production and newer ones for automobile propulsion by fuel cells have revived. Extensive fuel cell research throughout the world did not, however, fulfil these expectations. This research was not without value since a fundamental knowledge of the function of galvanic cells in general was obtained helping to improve the properties (the power, lifetime, weight and volume) of commonly used cells. Electrode electrochemistry has contributed a great deal to progress in technology but a technological revolution can hardly be expected in this field. It appears that future electrochemical research will be oriented primarily towards the subject of the next chapter, membranes.

Chapter 3
Membranes

Basic Features

The existence of living cells was first detected by the English physicist R. Hooke who in 1665 drew a sketch of cork network resembling a honeycomb. The Dutch lens-grinder, A. Leevenhoek (1628–1723), to whom the invention of the microscope is often ascribed, discovered the cells of erythrocytes, bacteria, infusoria, and spermia. In 1848 du Bois-Reymond suggested that the surface of a living cell has properties similar to an electrode of a galvanic cell and at the turn of the century W. Ostwald, W. Nernst, and J. Bernstein predicted that living cells are enclosed by a semipermeable membrane with characteristic electric properties. These were unfounded assumptions before Gorter and Grendel (1925) spread the lipids from erythrocytes of various origin to monolayers at the water/air interface and found that they occupy an area about twice as large as the surface of the cells. (The principal components of this residual matter are phospholipids, usually glycerol esterified by fatty acids, and phosphoric acid with further components— for details see p. 151.) This means that the outer sheath of the cell is formed by a bimolecular layer of phospholipids. Later it was shown that all cells are covered by such a thin membrane composing no more than two layers of molecules, electron microscopy supplying conclusive proof. With plants the situation is more complicated, because the surface of their cells, in addition to having a cell membrane immediately enclosing the cell, is reinforced by the cell wall which is a relatively thick layer formed by cellulose and other materials (Fig. 67).

The membranes envelop not only the cells, but also various corpuscles inside the cell termed organelles. They include, for example, the cell nucleus, the endoplasmatic reticulum, the mitochondria (small structures in charge of cell breathing, see Figs. 68 and 69) and, in green plants, the chloroplasts consisting of still smaller corpuscles (Fig. 70), termed thylakoids, which provide for photosynthesis. The membrane is the site of the basic processes of life. As shown in Fig. 71, membrane processes include both the utilization of solar energy through photosynthesis by means of which sugars are synthesized from water and carbon dioxide, and exploitation of the sugars as a fuel in cell breathing.

While the importance of cell membranes was recognized as early as at the end of the nineteenth century,* the search for models of these membranes started at

---

* The German physicochemist Wilhelm Ostwald who coined the term 'semipermeable membrane' (a thin layer separating two solutions and permitting various ions to permeate with different facility), wrote in 1891: 'Not only electric currents in muscles and nerves but particularly even the mysterious effects occurring with electric fish can be explained by means of the properties of semipermeable membranes.'

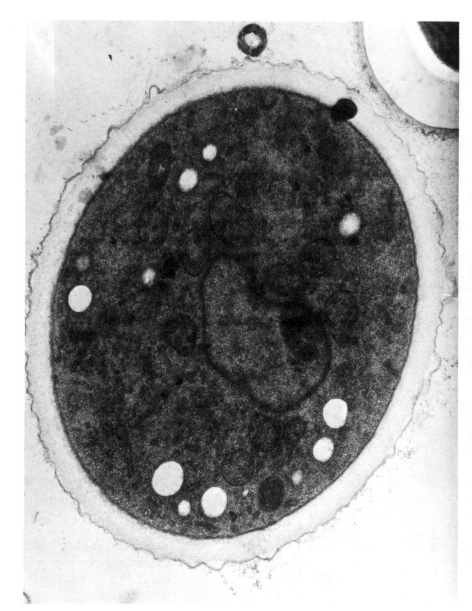

Fig. 67. Electron micrograph of a section from a yeast cell. The outer envelope is the cell wall. The inner double line is the cytoplasmatic membrane. Magnification $32,400 \times$. By courtesy of J. Ludvík

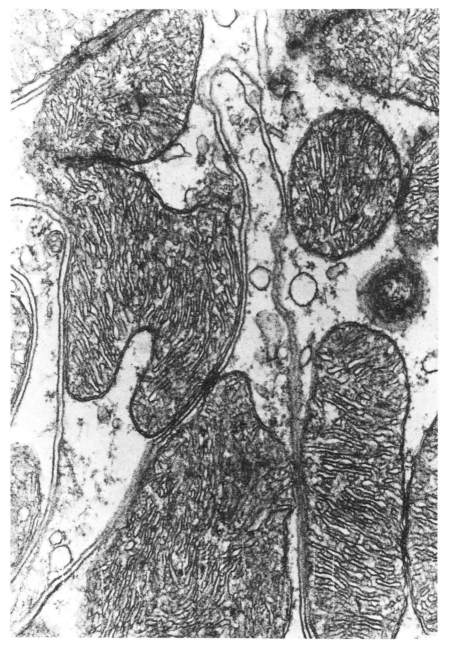

Fig. 68. Electron micrograph of mitochondria from rat kidney. On the ultra-thin section the structure of the cytoplasmatic membrane as well as of the mitochondria, including their membranes and the cristae, can be observed. Magnification 45,000'× . By courtesy of J. Ludvík

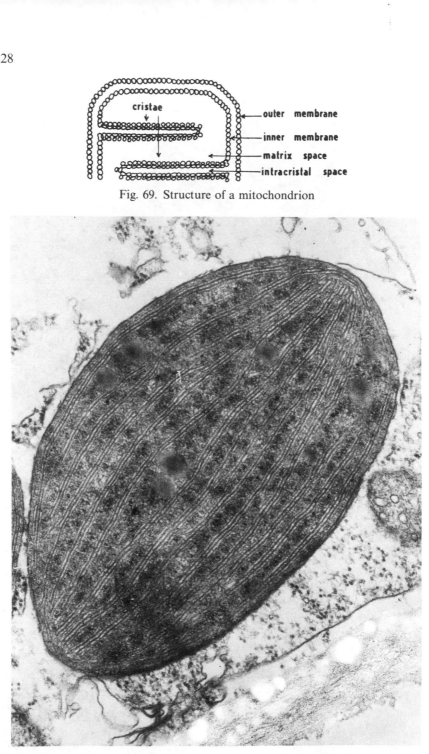

Fig. 69. Structure of a mitochondrion

Fig. 70. Electron micrograph of a section of a chloroplast from the green alga, Vaucheria. The structure of the basic units of the chloroplast, thylakoids, is easily perceptible. Magnification 33,600 ×. By courtesy of J. Ludvík

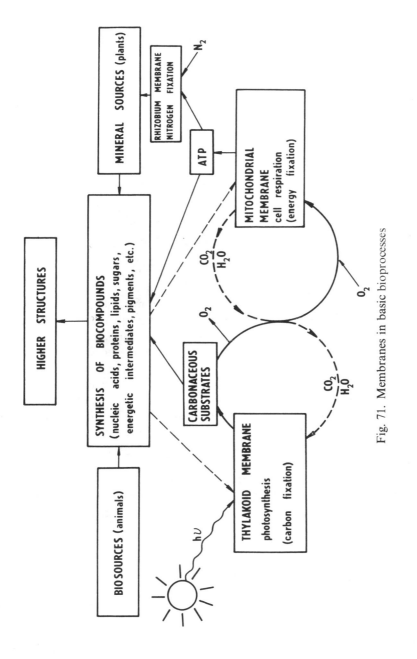

Fig. 71. Membranes in basic bioprocesses

the same time. The artificial membranes differed in many aspects (in thickness, for example) from cell membranes but they exhibit a number of interesting properties. In the course of the twentieth century a new branch of science was established concerned with the physical, chemical and biological rules that membranes obey and with membrane application in many regions of everyday life. Sometimes this interdisciplinary study is called *membranology*.[1]

How is a membrane defined? It is a layer which separates two solutions, differs in chemical composition from both of them, forms a sharp boundary towards both these solutions, and exhibits different permeability for individual components of the solution. The liquid junction of two electrolyte solutions described on p. 57 is of course not a membrane. A porous body (a porous *diaphragm*), where one solution passes into the other one, exerts no influence on the rate of transport of individual components of the solutions and only prevents mixing by convection. We shall be interested in membranes through which the *ions* penetrate with different ease, termed *electrochemical membranes*. In order to understand at least some of the phenomena occurring at biological membranes, very simple models, which are seemingly quite distant from living nature, must be used.

This is a difficult task when we consider the extreme complexity of the biological membrane. Fig. 72 shows how diverse particles may be embedded in a membrane (this is only a set of selected examples, a mere simplification of the actual complicated membrane structure). This cheerful picture was called the Chop-Suey model of a biological membrane by its author H. Ti Tien. (The favourite Cantonese Chop-suey type dishes contain a mixture of various ingredients starting with noodles and bamboo shoots and ending with fish and shrimps.) It would seem that the task of deciphering the intricacies of a system of such complexity indeed exceeds human powers. However, if we gave up we would lose the hope of elucidating basic processes in organisms. It appears that efforts tending to extreme simplification by isolation and artificial synthesis of limited sections and structural elements of these complicated systems lead to hopeful results.

Here we shall only be concerned with phenomena connected with ion transport across membranes and with electric phenomena occurring at membranes. However, this limitation is not as great as it seems because a considerable part of membrane phenomena is linked with electric currents and potentials.

Reference

1. Biological and artificial membranes are the main subject of the following journals: *Journal of Membrane Science, Journal of Membrane Biology and Biochimica et Biophysica Acta, Section Biomembranes*. Many papers dealing with membrane problems may be found in the *Journal of Bioenergetics and Biomembranes, The Biophysical Journal, Biophysical Chemistry, Journal of Colloid and Interfacial Chemistry, Journal of Electroanalytical Chemistry and Interfacial Electrochemistry, Section Bioelectrochemistry and Bioenergetics, Archives of Biochemistry and Biophysics, Biophysics of Structure and Mechanism, European Journal of Biochemistry, Accounts of Chemical Research*, etc. Several periodical or semi-periodical monograph series are

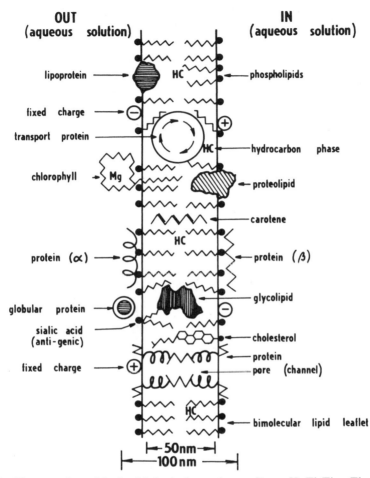

Fig. 72. 'Chop-suey' model of a biological membrane. From H. Ti Tien The scheme illustrates the variety of systems embedded in the membrane (this does not mean that they are present in a single membrane at the same time). The inside of the membrane is lipophilic, built of alkyl ('hydrocarbon') units. Reprinted from Ref. 1, p. 158, by courtesy of Marcel Dekker, Inc.

published containing reviews on various aspects of membranology, e.g. *Quarterly Reviews of Biophysics, Annual Reviews of Biophysics and Bioengineering, Cell Surface Reviews* (North-Holland, Amsterdam), *Membrane Transport in Biology* (Springer, Berlin), *Current Topics in Membranes and Membrane Transport* (Academic Press, New York), *Progress in Surface and Membrane Science* (Academic Press, New York), etc. Biological and other membranes are the subject of a number of monographs, for example, *Membranes* (G. Eisenman, ed.), 3 volumes, M. Dekker, New York, 1972–1975, *Membranes and Ion Transport* (E. E. Bittar, ed.), 3 volumes, John Wiley and Sons, London, 1970–71, A Kotyk and K. Janáček, *Cell Membrane Transport*, 2nd edn, Plenum, New York, 1975, *Membranes, Ions and Impulses* (J. W. Moore, ed.), Plenum, New York, 1976, *Membrane Transport Processes*, 3 volumes, Raven Press, New York, 1978–1979, *Structure and Function of Energy Transducing Membranes* (K. van Dam and B. F. van Gelder, eds.), Elsevier, Amsterdam, 1977, etc.

Ion-exchanging Membranes

Consider a piece of a cation-exchanging membrane (which is commercially available) placed in a cell (Fig. 73) so that it completely separates compartments 1 and 2. The cation-exchanging membrane is made of a copolymer of polyvinylchloride and sulphonated polystyrene. This material has a porous structure and the sulphonate groups—SO_3^- are fixed on the walls of the pores (Fig. 74). Compartments 1 and 2 are filled with a solution of potassium chloride with different concentration. The ratio of the concentrations is kept constant, for example $c_2/c_1 = 10$. In both compartments are placed identical saturated

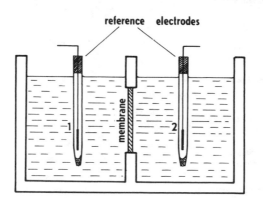

Fig. 73. A cell for the study of the membrane potential

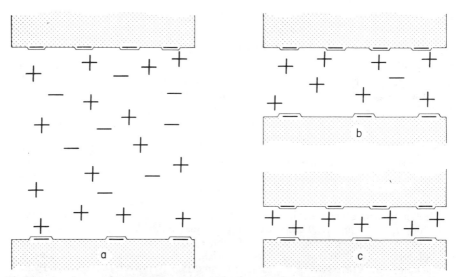

Fig. 74. A scheme of a membrane pore with negative fixed charge on its surface. a, wide pore with appreciable concentration of coions; b, the gegenions prevail in a narrower pore; c, the very narrow pore is completely filled with gegenions. By courtesy of K. Sollner

calomel electrodes which have a negligible liquid junction potential at the boundary with other aqueous electrolyte solutions (p. 57). The potential difference between solutions 2 and 1 measured using these reference electrodes is termed the membrane potential $\Delta\varphi_M = \varphi(2) - \varphi(1)$. At very low concentrations of potassium chloride the membrane potential is approximately $-59\,\text{mV}$. When the concentration of KCl increases (the ratio c_2/c_1 being kept constant) the absolute value of the potential difference decreases and, finally, at high KCl concentrations it approaches zero.

This behaviour of the membrane is closely linked to its structure. The fixed anions in the pores which are filled with water exert an electrostatic influence on the ions of the electrolyte so that a diffuse electric double-layer is formed in the pores (see Fig. 74). The effective thickness of the diffuse double-layer strongly depends on the concentration of the electrolyte (see p. 112). Thus, at two electrolyte concentrations the pores are filled only by the gegenions (see p. 42) which, in the present case, are positively charged. The coions are completely absent from the pores. Only the cations mediate the electric contact between both solutions (this may be directly demonstrated using radioactively labelled ions). At equilibrium the electrochemical potentials of the cations are equal so that for a dilute solution

$$\Delta\varphi_M = -(RT/F)\ln(c_2/c_1) \tag{3.1}$$

This potential difference is termed the *Nernst potential*. With an anion-exchanging membrane (containing, for example, fixed $-\text{N(CH}_3)_3^+$ groups) with otherwise identical properties (average pore thickness, number of ionized group per unit area of the pore surface) the Nernst potential has the same value but the opposite sign. The membrane which permits only ions of one sign to permeate is termed *permselective*.

At higher electrolyte concentrations the effective thickness of the diffuse double-layer becomes comparable with the pore radius and even smaller. The coions penetrate into the pores and the membrane acquires the properties of a liquid junction influenced by the presence of fixed ions on the walls of the pores. Finally, at rather high electrolyte concentrations, the double-layer shrinks to a simple Helmholtz structure (p. 112) and the inside of the pore is an electroneutral mixture of gegenions and coions. Thus, the membrane has degenerated to a mere diaphragm with an internal liquid junction (the zero final value of the membrane potential is due to the almost equal mobilities of K^+ and Cl^-).

J. Bernstein in 1902 and L. Michaelis in 1925 suggested describing the behaviour of this membrane type by analogy with equation (1.46)

$$\Delta\varphi_M = -(RT/F)(\tau_B - \tau_A)\ln(c_2/c_1) \tag{3.2}$$

where τ_B and τ_A are the transport numbers of the monovalent cation and of the monovalent anion, respectively, which depend on the concentration of the electrolytes in contact with the membrane, on the sign of the fixed ions, and on the membrane structure (for example, for a cation-exchanging membrane and rather dilute solutions $\tau_B \to 1$ and $\tau_A \to 0$).

The distribution of the ions in a porous membrane is shown in Fig. 75. A more exact theory for the membrane potential was deduced by T. Teorell and by K. H. Meyer and J.-F. Sievers.[1]

The ion-exchanging properties of porous membranes display interesting effects under the flow of an electric current across the membrane. At low concentrations the membrane only permits passage of ions of opposite sign to the fixed ionic groups, which can be utilized in preparative electrolysis. The membrane acts as a separator preventing the ionic products formed at one electrode from reaching the other electrode. At higher concentrations it loses its discriminating properties but keeps the gases at the electrodes where they should react (see p. 87).

A very attaactive application of membrane electrolysis is *electrodialysis*. By means of dialysis the high molecular and low molecular components of a solution are separated in an experimental arrangement where a membrane (not necessarily an electrochemical membrane) is placed between the solution and pure water. Only low-molecular species can permeate across the membrane while the high-molecular components remain in the original solution. Electrodialysis is based on the ability of the porous ion-exchanging membrane to permit permeation of ions of definite charge sign, either cations or anions. The main features of this method will now be shown using the example of desalination of drinking water, which is at present the most important application of electrodialysis. Fig. 76 depicts an electrodialytic cell consisting of parallel alternating cation- and anion-exchanging membranes.

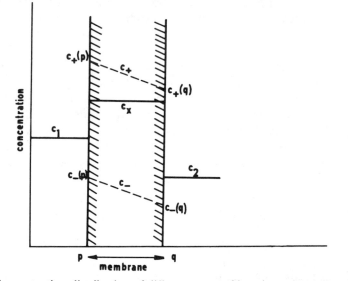

Fig. 75. Concentration distribution of different types of ions in a porous membrane. c_x denotes the concentration of fixed anions, c_+ and c_- the concentrations of mobile ions in the membrane, and c_1 and c_2 the electrolyte concentrations in adjacent solutions. At interfaces p and q Donnan-type equilibria are established, while inside the membrane a potential difference of the liquid junction type is formed. According to K. Sollner

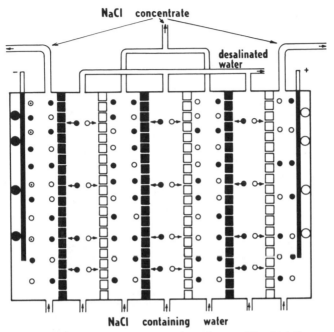

NaCl concentrate

desalinated water

NaCl containing water

Fig. 76. An electrodialytic cell for the desalination of water. The thick line segment with a minus sign is the cathode where hydrogen bubbles (●) are evolved. The anode is represented by the thick line segment with a plus sign. ○ represents a chlorine bubble. Small full circles are sodium ions, small void circles chlorine ions. ■ ■ ■ ■ is the cation-exchanging membrane and □ □ □ □ the anion-exchanging membrane. According to Fischbeck

Water containing NaCl is brought into the cell from below. When an electric current flows across the membrane system one compartment will be enriched with NaCl while desalinated water will remain in the other one. Desalinated water is then pumped from the upper part of the cell for direct use while the more concentrated NaCl solution is drained for further processing.[2]

Two *electrokinetic phenomena* (cf. p. 114) occur in porous membrane systems. When an electric current flows across the membrane it sets the charges in the diffuse double-layer in motion (in a cation-exchanger membrane in the direction of the current and vice versa). Simultaneously, the solvent is dragged with the ions so that a kind of osmosis—electro-osmotic flow—ensues. The more dilute the electrolyte solution the thicker is the diffuse double-layer region, so that the electro-osmotic flow increases with decreasing electrolyte concentration. This flow can be stopped by a pressure difference—the *electro-osmotic pressure*—between the two sides of the membrane (Fig. 77).

The inverse phenomenon to electro-osmotic flow appears when the electrolyte solution is pressed through the membrane in the absence of an electric current. The charge in the electrical diffuse double-layers in the pores of the membrane cannot of course be dragged with the liquid so that an electric field is formed in the

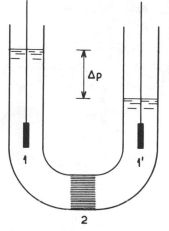

Fig. 77. Electro-osmotic pressure Δp prevents the electrolyte solution from crossing the porous ion-exchanging membrane

pores in the direction parallel to their axis. The final result is the electric potential difference—the *electrokinetic potential*—between the two sides of the membrane.

Besides porous membranes discriminating among the ions according to their charge, there are membranes which permit only small ions to permeate while macromolecular ions are prevented from crossing the membrane. Sir Frederick Donnan investigated the case when compartment 1 (Fig. 73) contains a salt consisting of cation B^+ and anion A^-, which can both permeate across the membrane, while compartment 2 contains a salt of the same cation B^+ with macromolecular cation C^-. Because this system is not in equilibrium, electrolyte BA will diffuse from 1 into 2. Equilibrium will be established as soon as the electrochemical potentials of the permeating ions B^+ in both the compartments are equal

$$\tilde{\mu}_B(1) = \tilde{\mu}_B(2)$$
$$\tilde{\mu}_A(1) = \tilde{\mu}_A(2)$$

These relationships yield the equation of the *Donnan equilibrium*

$$a_B(1)a_A(1) = a_B(2)a_A(2) \tag{3.3}$$

where the a_B's and a_A's are the activities of B^+ and A^-, respectively.

In addition to porous ion-exchanging membranes there are a number of other artificial membrane types, particularly the thick compact membranes, which may be either solid or liquid, and very thin bilayer lipid membranes (consisting of two contacting monomolecular layers).

References

1. T. Teorell, *Trans. Faraday Soc.*, **33** (1937) 1054.
   K. H. Meyer and J. F. Sievers, *Helv. Chim. Acta*, **19** (1936) 649, 665.
   R. Schlögl, *Stofftransport durch Membranen*, Steinkopff, Dermstadt, 1964.

D. Meares, J. F. Thain, and D. G. Dawson, 'Transport across ion-exchange resin membranes: The functional model transport', in: *Membranes*, Vol 1 (G. Eiseman, ed.), Marcel Dekker, New York, 1972, p. 55.

For the historical development of the electrochemistry of porous membranes, see K. Sollner, 'The early development of the electrochemistry of polymer membranes', in: *Charged Gells and Membranes*, Vol. 1 (E. Selegny, ed.), Reidel, Dordrecht, 1976.

2. K. S. Spiegler, *Principles of Desalination*, Academic Press, New York, 1966.

R. E. Lacey and S. Loeb, *Industrial Processes with Membranes*, Wiley–Interscience, New York, 1972.

Ion-selective Electrodes

In this section we shall be concerned with thick compact membranes. Compact membranes have either fixed ion-exchanging sites (membranes made of solid or glassy materials) or mobile ion-exchanging sites (liquid membranes). This sort of membrane is a basic component of important analytical devices, namely ion-selective electrodes.[1] The properties of a compact glassy membrane will be demonstrated using a simple glass electrode.[2] A very thin bubble is blown from a tube of sodium–calcium glass (for example, glass Corning 015 with $22\,wt\%$ Na_2O, $6\,wt\%$ CaO, and $72\,wt\%$ SiO_2, final glass thickness about 0.1 mm). The bulb is filled with a 0.1 M HCl. A silver–silver chloride electrode is then mounted in the bulb to dip into the inner solution of this glass electrode (Fig. 78). The whole device is then soaked in acetate buffer, $pH = 4.73$, for two days. To investigate the response of this glass electrode to solutions over the whole pH-range, the electrode will be immersed in different buffer solutions together with a saturated calomel electrode. (For solutions with a pH above 12 or below 2,

Fig. 78. A glass electrode. 1, glass membrane; 2, silver–silver chloride electrode; 3, 0.1 M HCl; 4, shielded lead, 5, connector to the electronic voltmeter (a 'pH meter')

138

NaOH or HCl, respectively, is used at different concentrations.) Thus, the galvanic cell

$$\text{Hg} \left| \text{HgCl} \right| \text{sat. KCl} \left| \text{buffer solution} \right\| 0.1 \text{ M HCl} \left| \text{AgCl} \right| \text{Ag}$$

glass
membrane (3.4)

is formed, the EMF of which, called the 'potential of the glass electrode' is plotted as a function of the *pH* in Fig. 79. Up to *pH* $\approx$ 10 the plot is linear with a slope of 59.1 mV per *pH* unit (at 25°C),

$$E_{gl} = E_{0,gl} - 59.1pH \tag{3.5}$$

where $E_{0,gl}$ is a constant. This slope, which can be expressed as $2.3\,RT/F$, is characteristic of dependences of the Nernst-type equation and is thus termed the Nernstian slope. However, at large *pH* values a strong deviation of the independence from linearity is observed; finally, the EMF is independent of *pH*. This phenomenon is called the glass electrode alkaline, or sodium error.

The EMF of a cell consisting of an electrochemical membrane separating two solutions with immersed reference electrodes (for example (3.4)) is given by equation

$$E = E_2 - E_1 + \Delta\varphi_M$$

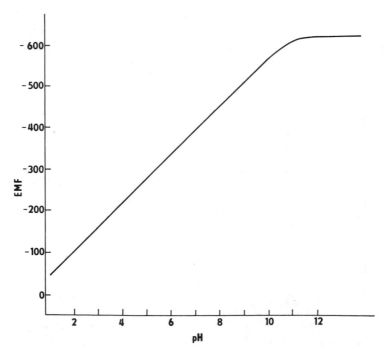

Fig. 79. The dependence of the EMF of the glass electrode–saturated calomel electrode cell on the pH of the analysed solution

where E_2 is the potential of the right-hand electrode and E_1 of the left-hand electrode. It follows from the dependence in Fig. 79 that

$$\Delta\varphi_M = (RT/F)\ln\left(a_{H_3O^+}(1) + K_1\right) + K_2 \tag{3.6}$$

where $K_2 = -(RT/F)\ln a_{H_3O^+}(2)$.

Varying the concentration of sodium ions at constant pH yields

$$K_1 = K_3 a_{Na^+}$$

Combining this equation with (3.6) gives

$$\Delta\varphi_M = (RT/F)\ln\left[(a_{H_3O^+}(1) + K_3 a_{Na^+})/a_{H_3O}(2)\right] \tag{3.7}$$

Thus, for $a_{H_3O^+}(1) \gg K_3 a_{Na^+}$ the membrane potential is reminiscent of the Nernst potential (3.1)

$$\Delta\varphi_M = (RT/F)\ln a_{H_3O^+}(1)/a_{H_3O^+}(2) \tag{3.8}$$

corresponding to a situation where only hydronium ions can cross the membrane. In rather alkaline solutions the membrane reacts only to sodium ions (to a small degree also to ions of other alkali metals)

$$\Delta\varphi_M = (RT/F)\ln\left[a_{Na^+}(1)/a_{H_3O^+}(2)\right] + \text{const.}$$

Thus, the response of the electrode, which originally reacted to hydronium ions, changes in alkaline solution to sensitivity for sodium ions. As pointed out by B. P. Nikolsky almost half a century ago, this behaviour can be explained by the ion-exchanging properties of the glass. In aqueous solution the surface layer of the glass becomes hydrated and in acidic, neutral, and weakly alkaline solutions hydrogen ions replace the sodium ions. The hydrogen ions in the glass and in the solution then determine the electrical potential difference at the glass/solution interface. At higher pH-values, the activity of the hydronium ions in the solution decreases to such a degree that the hydrogen ions are no longer retained in the glass but are replaced by sodium ions.

In order to explain this behaviour quantitatively, assume that the membrane potential can be divided into three components: the Nernst potential at the solution(1)/membrane $\Delta\varphi_p$ interface; the potential inside the membrane $\Delta\varphi_D$ (which is a diffusion potential); and the Nernst potential at the membrane/solution(2) interface $\Delta\varphi_q$,

$$\Delta\varphi_M = \Delta\varphi_p + \Delta\varphi_D + \Delta\varphi_q \tag{3.9}$$

For the sake of simplicity, the diffusion potential $\Delta\varphi_D$ will be neglected. The Nernst potentials $\Delta\varphi_p$ and $\Delta\varphi_q$ are given by the equations

$$\Delta\varphi_p = \varphi(p) - \varphi(1) = -(RT/F)\ln\left[C_{H^+}(p)/a_{H_3O^+}(1)\right] \tag{3.10}$$

$$\Delta\varphi_q = \varphi(2) - \varphi(q) = (RT/F)\ln\left[C_{H^+}(q)/a_{H_3O^+}(2)\right] \tag{3.11}$$

The points inside the membrane close to solution (1) are denoted by p and these close to solution (2) by q. Since very little is known about the activity coefficients of ions in the glass phase, concentrations are used instead. Thus, C_{H^+} denotes the concentration of hydrogen ions in the glass which, except for alkaline solutions, is equal to the concentration of fixed sites for hydrogen ions in the structure of the glass C_S. Therefore $C_{H^+}(p) = C_{H^+}(q) = C_S$. Combination with equations (3.9), (3.10), and (3.11) yields equation (3.8). In alkaline solution the exchange reaction

$$Na^+(w) + H^+(glass) + H_2O(w) \rightleftarrows Na^+(glass) + H_3O^+(w)$$

must be considered (w = water). The equilibrium constant of this reaction is given by the equation

$$K_{exch} = \frac{C_{Na^+} \cdot a_{H_3O^+}}{a_{Na^+} \cdot C_{H^+}} \qquad (3.12)$$

(the constant activity of water is included in K_{exch}).

Furthermore, assume that

$$C_{Na^+} + C_{H^+} = C_S \qquad (3.13)$$

Substitution from (3.12) and (3.13) into (3.10) yields

$$\Delta\varphi_p = (RT/F)\ln\left[a_{H_3O^+}(1) + K_{exch}a_{Na^+}(1)/C_S\right] \qquad (3.14)$$

As $C_H(q) = C_S$, because the internal solution of the glass electrode is acidic, the final result, obtained by combining (3.9), (3.11), and (3.14), is identical to the empirical equation (3.7), where $K_3 \equiv K_{exch}$. Equations of the type (3.7) (Nikolsky equations) have wide application in the field of ion-selective electrodes. They are usually written in a form containing only terms referring to the solution considered (1)

$$E_{ISE} = E_0 + (RT/zF)\ln(a_D + K_{DI}a_I^{z_D/z_I}) \qquad (3.15)$$

where a_D is the activity of the ion which is to be determined (determinand), a_I is the activity of the interfering ion (interferant), z_D and z_I are their charge numbers, and K_{DI} is the selectivity coefficient for ion I with respect to ion D.

For common glass electrode materials, the selectivity coefficient $K_{H_3O^+,Na^+}$ has a value of about 10^{-11}. This type of glass is suitable for daily laboratory use and particularly for biological and medical applications where measurements above $pH10$ are rare. For practical measurements the glass electrode must first be calibrated. The EMF of cell (3.4) is determined for a number of standard buffer solutions (see, for example, Table 4) and in this way the value E_0 is evaluated for the given electrode. Then, the pH of an analysed solution can be determined according to equation (3.5).

Glass electrodes with suitable glass composition ·can be used for pH determination in quite alkaline media, for example when Li_2O is used instead of Na_2O. Common glasses consist of monovalent and divalent metal oxides combined with SiO_2. When the divalent metal oxide, for example CaO, is replaced with a trivalent metal oxide such as Al_2O_3, the sodium error increases considerably, so that the glass electrode functions in a wide pH-range as a sodium selective electrode (i.e. $K_{H_3O^+,Na^+}$ markedly increases). For example, an electrode

The electrode response to ammonia is a measure of the urea content in the solution.
made of glass containing 11 % Na_2O, 18 % Al_2O_3, and 71 % SiO_2 indicates the sodium ion activity according to the equation

$$E_{gl} = E_{0,gl} + (RT/F)\ln a_{Na^+}.$$

This property is preserved to *pH* values as low as 6 while the selectivity towards potassium ions is rather low. Thus this type of the glass electrode can be used for sodium ion determination, even in more than 1000 time excess of potassium ions.

Devices based on the glass electrode can be used to determine certain gases present in a gaseous or liquid phase. Such a gas probe (Fig. 80) consists of a glass electrode covered by a thin film made of a plastic material with very small pores which is hydrophobic, so that the solution cannot penetrate into the pores (cf. p. 121). A thin layer of an indifferent solution is present between the surface of the film and of the glass electrode and is in contact with a reference electrode. The gas (ammonia, carbon dioxide, etc.) permeates through the pores of the film and dissolves in the solution at the surface of the glass electrode. A definitive value of the *pH* corresponding to the equilibrium concentration of a gas such as ammonia fixes the potential of the glass electrode in this solution. The actual determination of the ammonia concentration is based, of course, on a calibration plot of the electrode potentials versus the known ammonia concentration in the solution analysed.

A similar arrangement is typical of some *enzyme electrodes*.[3] The basic component of these sensors is a hydrophilic polymer containing an enzyme which transforms the determinand into a substance which is sensed by the electrode. For example, the urea electrode consists of an ammonia probe (Fig. 80) where the porous film is covered with a nylon net (G. G. Guilbault who invented this sensor originally used a piece of a nylon stocking). A polyacrylate gel containing the enzyme urease is spread on this net. Urea penetrates from the analysed solution into the gel layer where it is decomposed according to the equation

$$CO(NH_2)_2 + H_2O \xrightarrow{\text{urease}} CO_2 + 2NH_3$$

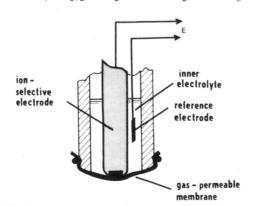

Fig. 80. A gas probe (manufactured by Orion Research)

142

Instead of potentiometric indication, some sensors are based on amperometric determination of the product produced by the enzymatic reactions. For example, an electrochemical sensor for determination of glucose in blood contains the glucose oxidase immobilized in polyacrylamide gel and a platinum electrode. The enzymatic reaction is

$$glucose + O_2 \xrightarrow[\text{oxidase}]{\text{glucose}} H_2O_2 + gluconic\ acid$$

The product, H_2O_2, is oxidized at the platinum electrode.

Instead of an enzyme-containing layer, a suspension of bacteria can sometimes be used, as shown in Fig. 81. In this bacterium electrode, glutamic acid is decomposed by bacteria and the concentration of the ammonia released is measured with an ammonia gas probe.[4]

Properties analogous to those of the glass electrode are exhibited by a number of ion-selective electrodes with fixed ion-exchanging sites. The membrane in these electrodes is a layer of solid electrolyte. The most important of this group is the fluoride ion-selective electrode, a single crystal of lanthanum fluoride functioning as the membrane. The only interferants in fluoride ion determination are hydroxide ions. This electrode has many applications—from fluoride determination in waters to analysis of saliva and tooth enamel.

Ion-selective electrodes for halide ions are based on a mixture of silver halide with silver sulphide. Silver sulphide alone forms the membrane of an ion-selective electrode for determination of sulphide and silver ions. Ion-selective electrodes are produced commercially by a number of companies throughout the world.

Before dealing with liquid-membrane ion-selective electrodes, the electrochemical properties of the interface of two immiscible electrolyte solutions (ITIES), which is the site of basic processes occurring in this type of electrode, will be considered.

References

1. J. Koryta, *Ion-selective Electrodes*, Cambridge University Press, 1975.
   H. Freiser, Ed., *Analytical Applications of Ion-selective Electrodes*, Plenum, New York, Vol. I, 1978, Vol. II, 1980.

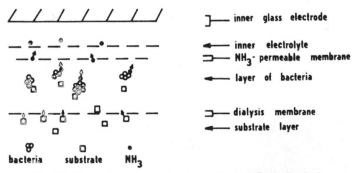

Fig. 81. Bacterium electrode. According to G. A. Rechnitz

2. G. Eisenman, Ed., *Glass Electrodes for Hydrogen and Other Cations: Principles and Practice*, Marcel Dekker, New York, 1967.
3. G. G. Guilbault, 'Use of enzyme electrodes in biomedical investigations', in: *Medical and Biological Applications of Electrochemical Devices* (J. Koryta, ed.), John Wiley & Sons, Chichester, 1980.
4. P. D'Orazio, M. E. Meyerhoff, and G. A. Rechnitz, *Anal. Chem.*, **50** (1978) 1531.

Interface between Two Immiscible Electrolyte Solutions

Two liquids with low miscibility, for example water and tetrachloromethane CCl_4, will be examined. The mutual solubility of these liquids is very low. Iodine is dissolved in CCl_4 to give a concentration of $10^{-3}\,mol/dm^3$. When the resulting solution is shaken with the same volume of water some of the iodine enters the aqueous phase. Analysis of the resulting content of iodine in the two solvents reveals that the ratio of its concentrations in tetrachloromethane and water is 85 at 25°C. If the initial concentration of iodine in CCl_4 is increased to $10^{-2}\,mol/dm^3$, the resultant concentration ratio after shaking with water is the same. The distribution equilibrium of a non-ionized substance between two phases β and α is characterized by an equilibrium constant called the distribution coefficient

$$K_i^{\beta,\alpha} = \frac{a_i(\beta)}{a_i(\alpha)} \tag{3.16}$$

In this equation β denotes CCl_4, α water, and a_i the iodine activity. Thermodynamic interpretation of this equation is simple. At equilibrium the chemical potentials of the dissolved substance must be the same in both phases:

$$\mu_i(\alpha) = \mu_i(\beta)$$
$$\mu_i(\alpha) + RT\ln a_i(\alpha) = \mu_i^0(\beta) + RT\ln a_i(\beta) \tag{3.17}$$

Equation (3.16) is a direct consequence of equation (3.17). It holds for the distribution coefficient that

$$K_i^{\beta,\alpha} = \exp\left[-(\mu_i^0(\beta) - \mu_i^0(\alpha))/RT\right]$$

The difference in standard chemical potentials of a substance in phases β and α is termed the standard Gibbs energy of transfer of substance i from phase α to phase β

$$\Delta G_{tr,i}^{0,\alpha\rightarrow\beta} = \mu_i^0(\beta) - \mu_i^0(\alpha) = -RT\ln K_i^{\beta,\alpha} \tag{3.18}$$

Why does iodine prefer tetrachloromethane to water at equilibrium? The reason is that a stronger solvation of iodine can be found in the non-aqueous solvent than in water. The Gibbs transfer energy is given by the difference in the standard solvation energies

$$\Delta G_{tr,i}^{0,\alpha\rightarrow\beta} = \Delta G_{s,i}^{0,\beta} - \Delta G_{s,i}^{0,\alpha}$$

Consider the distribution of an electrolyte between CCl_4 and water, e.g. tetrabutylammonium iodide $(C_4H_9)_4NI$. This salt is soluble in both the solvents.

After shaking an aqueous solution of this substance with CCl_4 we find that the experiment was not chosen favourably. The tetrabutylammonium iodide passes almost completely into the non-aqueous phase. Nonetheless, the conductivity of this phase is negligible. The iodine anions and tetrabutylammonium cations are completely associated in ion pairs (because of the low permittivity of the solvent—cf. pp. 5 and 28). Consequently, another non-aqueous solvent that is immiscible with water must be chosen, for example nitrobenzene, where the salt is soluble and, at the same time, almost completely dissociated. The distribution coefficient is given by the equation

$$K_{BA}^{\beta,\alpha} = \frac{a_{BA}(\beta)}{a_{BA}(\alpha)} = \frac{a_{B^+}(\beta) \cdot a_{A^-}(\beta)}{a_{B^+}(\alpha) \cdot a_{A^-}(\alpha)} \tag{3.19}$$

Here B^+ denotes the terabutylammonium ion, A^- the iodide ion, α water, and β nitrobenzene. Relationship (3.18) is again valid between the distribution coefficient and the standard Gibbs energy of BA. Is an analogous relationship valid for the distribution equilibrium of the individual ions? If so, it would hold that

$$K_{B^+}^{\beta,\alpha} = \frac{a_{B^+}(\beta)}{a_{B^+}(\alpha)}$$

$$K_{A^-}^{\beta,\alpha} = \frac{a_{A^-}(\beta)}{a_{A^-}(\alpha)}$$

Unfortunately the situation is not that simple. Both the phases could be electrically charged. In fact, the individual phases in contact containing a dissolved electrolyte will almost always be charged because one ionic species can pass from phase α to phase β more easily than the other species and, consequently, an electric double-layer is formed at the interface. However, an individual ion cannot be transferred *from the bulk* of phase α *into the bulk* of phase β because the electroneutrality of these phases would be disturbed (which is not true when the electric double-layer is formed or its charge changed).

Thus, a relationship must be written for the distribution coefficient which would include the electric potential difference between the two phases $\Delta_\alpha^\beta \varphi = \varphi(\beta) - \varphi(\alpha)$. This relationship, which is based on the equality of the electrochemical potentials of the ion in the two phases, has the form

$$K_{B^+}^{\beta,\alpha} = \exp\left[-(\mu_{B^+}^0(\beta) - \mu_{B^+}^0(\alpha))/RT)\right]\exp\left[-F\Delta_\alpha^\beta\varphi(/RT)\right]$$
$$= \exp\left[-\Delta G_{tr,B^+}^{0,\alpha\to\beta}/(RT)\right]\exp\left[-F\Delta_\alpha^\beta\varphi/(RT)\right]$$
$$K_{A^-}^{\beta,\alpha} = \exp\left[-(\mu_{A^-}^0(\beta) - \mu_{A^-}^0(\alpha)/(RT)\right]\exp\left[F\Delta_\alpha^\beta\varphi/(RT)\right]$$
$$= \exp\left[-\Delta G_{tr,A^-}^{0,\alpha\to\beta}/(RT)\right]\exp\left(F\Delta_\alpha^\beta\varphi/(RT)\right)$$

Here $\Delta G_{tr,B^+}^{0,\alpha\to\beta}$ and $\Delta G_{tr,A^-}^{0,\alpha\to\beta}$ are the standard Gibbs energies for transfer of the ions B^+ and A^- from phase α and phase β. Obviously

$$K_{BA}^{\beta,\alpha} = K_{B^+}^{\beta,\alpha} \cdot K_{A^-}^{\beta,\alpha}$$

and

$$\Delta G_{tr,BA}^{0,\alpha\to\beta} = \Delta G_{tr,B+}^{0,\alpha\to\beta} + \Delta G_{tr,A-}^{0,\alpha\to\beta} \qquad (3.20)$$

Unfortunately, there are even further complications. The difference in the electric potentials between the liquids is not accessible to direct measurement. It cannot be assessed using identical reference electrodes attached to each phase by appropriate salt bridges—in at least one of the bridges an ITIES would be formed with unknown potential. At the same time the standard Gibbs energy of a salt cannot be separated into the standard Gibbs energies of the individual ions according to equation (3.20).

Both these difficulties could be eliminated by finding a cation–anion pair that are solvated to the same degree in any solvent. It is reasonable to assume[1] that such a pair is represented by tetraphenylarsonium tetraphenylborate (TPAs$^+$ TPB$^-$)

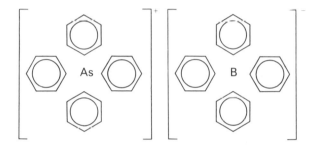

Both these ions are quite voluminous so that the charge of the central atom acts only electrostatically on the solvent dipoles. The interaction of the benzene rings with the solvent is the same for both ionic species. Thus, the same standard Gibbs energy of transfer from an arbitrary solvent α to another solvent β can be attributed to both ions

$$\Delta G_{tr,TPAs+}^{0,\alpha\to\beta} = \Delta G_{tr,TPB-}^{0,\alpha\to\beta}$$

On the basis of this assumption, the standard Gibbs transfer energy for an arbitrary ion can be determined. For example, when this quantity is to be determined for the iodide ion and two solvents, α and β, it is sufficient to measure the distribution coefficients for TPAsTPB and for TPAsI between these two solvents. The result for the iodide is given by the relationship

$$\Delta G_{tr,I-}^{0,\alpha\to\beta} = \Delta G_{tr,TPAsI}^{0,\alpha\to\beta} - \tfrac{1}{2}\Delta G_{tr,TPAsTPB}^{0,\alpha\to\beta}$$

Knowledge of the standard Gibbs energy of transfer from α to β for an individual ionic species helps to determine the electrical potential difference between these two phases. At equilibrium the electrochemical potentials, for example for a given species B$^+$, must be equal in both phases,

$$\tilde{\mu}_{B+}(\alpha) = \tilde{\mu}_{B+}(\beta)$$

146

It follows from this equation that

$$\Delta_\alpha^\beta \varphi = \Delta G_{tr,B^+}^{0,\alpha\to\beta}/F + (RT/F)\ln\left[a_{B^+}(\alpha)/a_{B^+}(\beta)\right] \tag{3.21}$$
$$= \Delta G_{tr,B^+}^{0,\alpha\to\beta}/F + (RT/F)\ln\left\{\gamma_{B^+}(\alpha)c_{B^+}(\alpha)/\left[\gamma_{B^+}(\beta)c_{B^+}(\beta)\right]\right\}$$

When both the solutions are dilute enough, activity coefficients $\gamma_{B^+}(\alpha)$ and $\gamma_{B^+}(\beta)$ will be determined using the Debye–Hückel Law (1.11).

The quantities $a_{B^+}(\alpha)$ and $a_{B^+}(\beta)$ are not independent since they are related together and to the activities of the corresponding anions by the equation for the distribution coefficient (3.19). It can be expected, therefore, that, for a given salt, $\Delta_\alpha^\beta \varphi$ will not depend on the amount of salt distributed between the phases. In fact, the potential difference $\Delta_\alpha^\beta \varphi$ depends on the anion activities according to an equation analogous to (3.21). Combining these two equations, and assuming that the activities of the cation and of the anion in a given solvent are equal, yields

$$\Delta_\alpha^\beta \varphi = \tfrac{1}{2}(\Delta G_{tr,A^-}^{0,\alpha\to\beta} - \Delta G_{tr,B^+}^{0,\alpha\to\beta})/F \tag{3.22}$$

The electrical potential difference governed by this equation, which is valid only when a single salt is distributed between two solvents, is termed the *distribution potential*.

The ITIES has further interesting properties, particularly when one of the solvents is water and the other is an organic liquid and when the aqueous phase contains a strongly hydrophilic electrolyte (e.g. LiCl) and the organic phase a strongly hydrophobic electrolyte like tetrabutylammonium tetraphenylborate. Then the properties of the ITIES become completely analogous to those of a polarizable electrode[2] (see 2.63).

The potential difference across the ITIES can be changed by introducing charges of opposite sign into each phase using electrodes immersed into these phases, most conveniently with a four-electrode potentiostatic system (for a three-electrode potentiostatic arrangement, see p. 90), as shown in Fig. 82.

When, for example, a semihydrophobic cation is present in the aqueous phase it can be 'forced' to cross the ITIES and to enter the organic phase. This process is

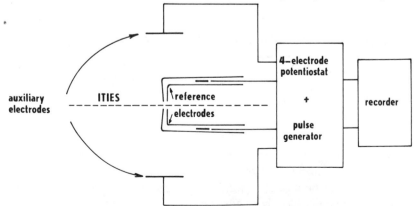

Fig. 82. A device (a voltage-clamp) for the polarization of the interface between two immiscible solutions by means of a four-electrode system

connected with the flow of an electrical current, where the strongly hydrophilic electrolyte in the aqueous phase and the strongly hydrophobic electrolyte in the organic phase act as indifferent electrolytes (cf. p. 64). The value of $\Delta_\alpha^\beta \varphi$ necessary for the transfer of a particular ion across the ITIES is connected with the value of the standard Gibbs transfer energy for that ion. This situation is completely analogous to electrolysis with metallic electrodes and the same methods of investigation as described on pp. 88–98 and 118–123 can be applied to electrolysis at ITIES.

References

1. A. J. Parker, *Electrochim. Acta*, **21** (1976) 671.
2. J. Koryta, *Electrochim. Acta*, **24** (1979) 293.
   J. Koryta and P. Vanýsek, 'Electrolysis at the interface of two immiscible electrolyte solutions', in: *Advances in Electrochemistry*, Vol. 12 (H. Gerischer and C. W. Tobias, eds.), Wiley–Interscience, New York, 1981.
   C. Gavach and F. Henry, *J. Electroanal. Chem.*, **54** (1974) 361.

Liquid–membrane Ion-selective Electrodes

Consider a non-aqueous phase containing a strongly hydrophobic anion A_2^- as an ion-exchanging ion and a counterion B^+ which is also present in the aqueous phase, together with a rather hydrophilic anion A_1^- (Fig. 83).

The potential difference across the ITIES is given by equation (3.21). Here the activity of B^+ in the organic phase is fixed by the presence of anion A_2^- which cannot cross the ITIES because of its hydrophobicity. Potential difference $\Delta_\alpha^\beta \varphi$ then depends only in the activity of B^+ in the aqueous phase

$$\Delta_\alpha^\beta \varphi = (RT/F) \ln a_{B^+}(\alpha) + \text{const.}$$

Thus, the ITIES can be used for sensing the determinand B^+. The organic phase then acts as a liquid membrane, usually fixed in a porous or polymeric structure. Another cation C^+ may act as an interferent which is soluble in the membrane as well as in the aqueous solution. Its interfering action is connected with its ability to expel the determinand ion from the organic phase through exchange reaction $B^+(\beta) + C^+(\alpha) = B^+(\alpha) + C^+(\beta)$. The equilibrium constant for this reaction is a function of the difference in the standard Gibbs transfer energies of the two ions between α and β. In this simple case the potential of the ion-selective electrode is again governed by the Nikolsky equation (3.15).

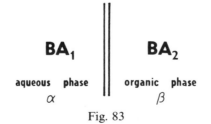

BA₁   **BA₂**

aqueous phase   organic phase
α β

Fig. 83

A useful arrangement for a liquid-membrane ion-selective electrode consists of a porous diaphragm (e.g. of porous Teflon or a Millipore filter) soaked with the organic ion-exchanging solution. A reservoir containing the organic solution is attached to the diaphragm to replace possible losses due to dissolution of the organic phase in the analysed aqueous solution (Fig. 84).

Another, at present very frequently used version is a plastic film made by mixing a solution of an appropriate polymer (particularly of the PVC type) in a suitable solvent with the ion-exchanging solution and drying this mixture on a sheet of plate glass. The ion-exchanging solution then acts as a plasticizer for the film (see Fig. 85). This method of preparation of a liquid membrane, introduced by G. J. Moody and J. D. R. Thomas in the present form, was originally used by K. Sollner in his pioneering work with collodion membranes.

Among liquid-membrane ion-selective electrodes, primarily the calcium, the nitrate, and the potassium electrodes should be mentioned. The ion-exchanging ion in the membrane of the electrode for calcium ion determination is the dialkyl phosphate ion, where the alkyl groups consist of chains of 8 to 10 carbon atoms. The function of liquid-membrane ion-selective electrodes is closely connected with the nature of the solvent. The calcium electrode, for example, senses calcium ions even in a hundred times excess of magnesium ions when the membrane solvent is dioctyl phenylphosphonate

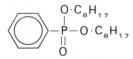

With 1-decanol as the membrane solvent, the electrode does not discriminate between calcium and magnesium and, therefore, is suitable for determination of

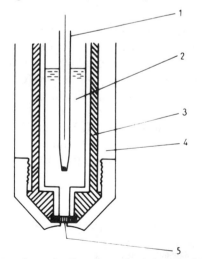

Fig. 84. A liquid-membrane ion-selective electrode. 1, inner reference electrode; 2, inner solution; 3, reservoir of the ion-exchanger solution; 4, plastic electrode body; 5, membrane prepared from a porous diaphragm soaked up with the ion-exchanger solution

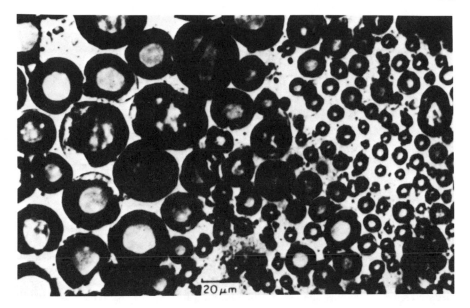

Fig. 85. Micrograph of a section from a PVC-plastic membrane. The pores (dark circles) contain the calcium salt of didecylphosphoric acid dissolved in dioctyl phenylphosphonate. By courtesy of G. H. Griffiths, G. J. Moody, and J. D. R. Thomas. Reproduced by permission of the Royal Society of Chemistry

water hardness. The nitrate electrode is based on the alkyl-substituted orthophenanthroline complex of divalent nickel or on a tetraalkylammonium cation with long alkyl chains.

One type of potassium electrode contains tetra(p-chlorophenyl) borate anion (cf. p. 145) as the ion-exchanging anion. The membrane solvent is o-nitrophenyloctyl ether. This electrode is used as a miniaturized sensor for potassium determination in excitable cells and tissues.[1] Ion-selective microelectrodes are either liquid-membrane electrodes, primarily for determination of potassium and also for determination of chloride, calcium, and lithium ions, or glass electrodes for hydrogen and sodium ions. The construction of ion-selective microelectrodes is similar to that of glass micropipettes (see p. 76). The membrane is an ion exchange solution filling the tip of the microelectrode. Usually it is necessary to attach a reference electrode as close as possible to the microelectrode. Otherwise the electrical potential difference between the positions of the orifices of the microelectrode and of the reference electrode resulting from the electrical field in the tissue would be added to the difference of the potentials of the microelectrode and of the reference electrode and an erroneous value of the determinand concentration would be obtained. The most useful geometry is the double-barrelled arrangement (see Fig. 86a). To improve the electric parameters of the microelectrode the longitudinal resistance of the ion-exchanging column is reduced by means of a second micropipette filled with KCl and plugged caxially into the column (Fig. 86b).

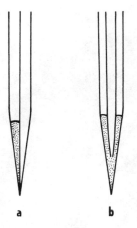

a   b

Fig. 86. Double barelled (a) and coaxial (b) ion-selective microelectrodes. The orifice of the ion-selective channel of the sensor is as close as possible to the orifice of the reference electrode channel, which is placed beside the ion-selective channel (a). In order to reduce the resistance of the ion-selective channel a micropipette filled with KCl solution is immersed in the ion-changing channel and acts as a connection to the inner reference electrode (b)

A very important group of liquid-membrane ion-selective electrodes is based on particular complex-formers, belonging among ion-carriers or ionophores (see p. 159). A potassium electrode of this type is based on a macrocyclic antibiotic, valinomycin (p. 160). A diagram of an ITIES with valinomycin V in the organic phase is shown in Fig. 87. The organic phase contains a complex of potassium with valinomycin, the same concentration of the hydrophobic cation, tetraphenylborate, and excess valinomycin, while the potassium salt of a hydrophilic anion is present in the aqueous phase. The potential difference $\Delta_\alpha^\beta \varphi$ is given by equation (3.21) ($K^+ = B^+$), where the activity of the potassium ion in the organic phase is expressed using the equation for complex formation

$$K_{VX} = a_{KV^+}/(a_{K^+} c_V)$$

where K_{VX} is the stability constant, c_V is the valinomycin concentration, and a_{KV^+} and a_{K^+} are the activities of the complexed and free potassium ions in the non-aqueous phase. Substitution from this equation into equation (3.21) yields

$$\Delta_\alpha^\beta \varphi = -\Delta G_{tr,K^+}^{0,\alpha \to \beta}/F + (RT/F)\ln\left[a_{K^+}(\alpha)K_{VX}c_V/a_{KV^+}(\beta)\right]$$
$$= (RT/F)\ln a_{K^+}(\alpha) + \text{const.}$$

Fig. 87

The potential difference at the ITIES depends only on the potassium activity in the aqueous phase because the other terms deal only with the situation in the membrane.

The potassium electrode based on valinomycin senses potassium in a ten thousandfold excess of sodium. The electrodes for calcium, sodium, and other alkali or alkaline earth metal ions are based on other, either naturally occurring or synthetic, ionophores.[2]

References

1. J. L. Walker, 'Single cell measurement with ion-selective electrodes', in: *Medical and Biological Applications of Electrochemical Devices* (J. Koryta, ed.), John Wiley & Sons, Chichester, 1980, p. 109.
   P. Hník, E. Syková, N. Kříž, and F. Vyskočil, 'Determination of ion activity changes in excitable tissues with ion-selective microelectrodes', in: *Medical and Biological Applications of Electrochemical Devices* (J. Koryta, ed.), John Wiley & Sons, Chichester, 1980, p. 129.
2. P. C. Meier, D. Ammann, W. E. Morf, and W. Simon, 'Liquid-membrane ion-selective electrodes and their biomedical applications', in: *Medical and Biological Applications of Electrochemical Devices* (J. Koryta, ed.), John Wiley & Sons, Chichester, 1980, p. 13.

Biological Membranes

So far, the properties of 'thick' membranes—between tenths and tens of millimetres—have been discussed. If the scale is increased 10^5 times, we enter the region of biological membranes. The discussion on p. 130 and Fig. 72 have already provided a certain picture of their diversified structure. The main components of biological membranes are lipids and proteins. A number of theories have been proposed for their arrangement in the membrane. These theories are not of a speculative nature; on the contrary, they are based on experimental research on the structure using methods such as electron microscopy, X-ray dispersion, study of monolayers of membrane components on water surfaces, electron spin resonance, and fluorescence spectroscopy of molecular probes implanted into the membrane, etc. Furthermore a membrane can be disintegrated into basic structural units and then resynthesized (this is called reconstitution of the membrane), etc. Fig. 88 shows how different structures can correspond to biological membranes with different functions. Not only various cells and organelles but even various sections of one membrane differ in their structure.

At least with some membranes the basic structural matrix is the bimolecular lipid layer—the 'bimolecular leaflet'. The lipids are oriented with their hydrophobic alkyl tails towards the inside of the membrane while their polar heads are directed toward the adjacent solution (Fig. 89).

The lipidic component of a membrane is represented primarily by phospholipids. The most important is the group of phosphoglycerides. The basic structure is phosphatidic acid

152

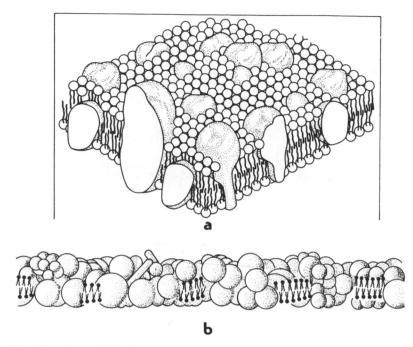

a

b

Fig. 88. Diversity of biological membrane structures. Many cytoplasmatic membranes have phospholipids as their major component and posses a 'liquid mosaic' structure (a). The irregular shaded structures are various proteins. In the mitochondrial membrane (b) proteins prevail over phospholipids. Adapted from S. J. Singer, G. L. Nicholson, and K. Dose

Fig. 89. A bilayer lipid membrane. Black circles represent polar heads and zig-zag lines alkyl chains

$$
\begin{array}{c}
\overset{\displaystyle O}{\underset{\displaystyle \|}{}} \\
\overset{\displaystyle O}{\underset{\displaystyle \|}{}}\quad H_2C-O-C-R \\
R'-C-O-CH \qquad \overset{\displaystyle O}{\underset{\displaystyle \|}{}} \\
H_2C-O-P-O-X \\
\underset{\displaystyle O^-}{|}
\end{array}
$$

(in the formula X = H), where R and R′ are alkyl or alkenyl long chain groups. Thus, glycerol is esterified with two higher fatty acids such as stearic, palmitic, oleic, linoleic, etc. acid and the orthophosphoric acid is bound on the remaining OH group. One of its acidic groups is esterified with ethanolamines, usually substituted on the nitrogen atom. For example, with binding choline (X = — $CH_2 \cdot CH_2 N(CH_3)_3^+$) the phospholipid lecithin is formed. In place of ethanolamines, other molecules such as serine, inositol, etc. can be bound to the phosphoric acid.

Instead of glycerol some phospholipids contain the ceramide,

$$
\begin{array}{l}
CH(OH)CH = CH(CH_2)_{12}CH_3 \\
| \\
CH-NHOCR \\
| \\
CH_2OH
\end{array}
$$

where R is a long-chain alkyl group. The ester of ceramide with phosphoric acid, also esterified with choline, is sphingomyeline. When ceramide is bound to a sugar, such as glucose or galactose through the β-glycosidic bond, cerebrosides are formed. Lipids also include cholesterol. In Table 9 the presence of individual types of lipids in various biological membranes is demonstrated.

Proteins play two roles in a biological membrane. Some of them simply support the texture of the membrane (structural proteins). Another, much more important, group of functional proteins participates directly in the membrane-processes (chemical reactions and metabolic transport). Proteins also differ in their position in the membrane—some of them are only adsorbed on its surface, while others (integral proteins) are embedded in its structure with only their polar sections reaching out or even penetrate across the membrane.

Several models have been proposed for the arrangement of lipids and proteins in a membrane. Some models (e.g. those proposed by J. F. Danielli and H. Davson, J. D. Robertson) are based on a bimolecular lipid leaflet with proteins adsorbed on both sides of the membrane in the form of an α-helix, a pleated leaflet, or a globule (see Fig. 90). A. A. Benson proposed a model comprising lipoprotein subunits for the thylakoid membrane. These subunits are held together by hydrophobic interactions between the alkyl groups of the lipids and the hydrophobic internal regions of the proteins. In some models micelles of lipids enclosed by proteins are considered, as proposed by F. Sjöstrand. Those formations can be converted to a pleated leaf structure. According to G. Vanderkooi and D. E. Green, the membrane is formed of protein globules diversified by lipids (see Fig. 90).

Table 9   Lipid components of some biological membranes (in %). After D. Chapman

Lipid	Myelinated sheath of nerve fibre	Erythrocyte	Mitochodrion	Azotobacter agilis	Escherichia coli
Cholesterol	25	25	5	—	—
Phosphatidylethanolamine	14	20	28	100	100
Phosphatidylserine	7	11	—	—	—
Phosphatidylcholine	11	23	48	—	—
Phosphatidylinositol	—	2	8	—	—
Phosphatidylglycerol	—	—	1	—	—
Sphingomyeline	6	18	—	—	—
Cerebroside	25	—	—	—	—
Ceramide	1	—	—	—	—
Other	12	2	11	—	—

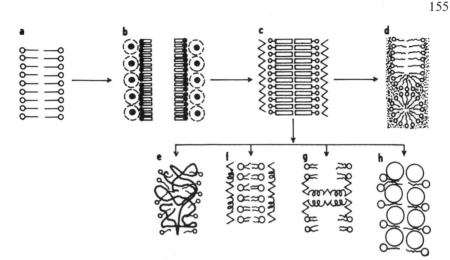

Fig. 90. Basic structures of biological membranes (from H. Ti Tien). a, a bimolecular leaflet of phospholipids (E. Gorter and F. Grendel); b, a bimolecular phospholipid leaflet with adsorbed protein globules (J. F. Danielli and H. Davson); c, a bimolecular phospholipid leaflet with adsorbed protein helix (J. D. Robertson); d, a model with a spherical lipid micelle (F. Sjöstrand); e, a lipoprotein model (A. A. Benson); f and g, models with different arrangement of hydrophobic lipoproteins in the membrane (J. Lenard and S. J. Singer); g, a model with a protein placed across the lipid film, h, model with protein subunits (D. E. Green). Reprinted from Ref. 1, p. 158 by courtesy of Marcel Dekker, Inc.

The variety of membrane structures is reflected in the relative content of proteins in the membrane, which is, for example, 18–25 % in the myelin of beef brain, 37–49 % in yeast, *Saccharomyces cerevisiae*, 53 % in human erythrocyte, 89 % in bird erythrocyte, 68 % in the bacterium *Micrococcus lysodeikticus*, and 85 % in rat liver.

Reference

J. B. Finean, R. Coleman and R. H. Michell, *Membranes and their Cellular Function*, Blackwell, Oxford, and Halsted Press, New York, 1979. See also p. 130–131.

Bilayer Lipid Membranes

So far the most successful model of a biological membrane is the bilayer lipid membrane (BLM).[1]

A cell containing a dilute electrolyte is divided into two compartments by a Teflon septum with a small hole (diameter about 1 mm). When a drop of a lipid in a suitable solvent, e.g. octane, is placed on the hole, a striking phenomenon is observed. The layer of the lipid solution gradually becomes thinner, rainbow interference colours appear on it and later also black spots and, finally, the whole layer becomes completely black (Fig. 91). The blackening marks the transition of the lipid layer from a multimolecular thickness to a bimolecular film, the bilayer lipid membrane (BLM) which is a non-reflecting optically black region. A BLM

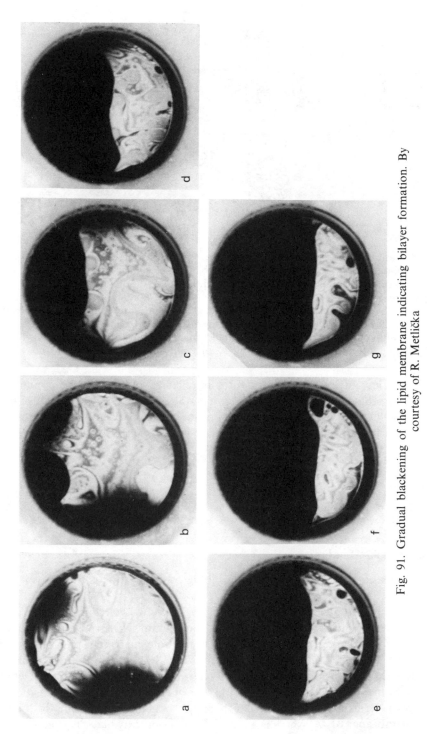

Fig. 91. Gradual blackening of the lipid membrane indicating bilayer formation. By courtesy of R. Metlička

in a solution was first prepared by P. Mueller, D. O. Rudin, H. Ti Tien, and W. C. Westcott in 1962, although similar phenomena with soap bubbles had already been observed by R. Hooke in 1672 and I. Newton in 1702. The attenuation of the layer is caused by interfacial forces together with the dissolution of the membrane solvent in the adjacent aqueous solution. The thick edge of the membrane is called the Plateau–Gibbs boundary (Fig. 92). The thickness of the membrane has been determined by means of low-angle light reflection, electric capacity measurement, and electron microscopy (a metal coating can, of course, result in artifacts). In Table 10 are listed the thicknesses of BLM prepared from various materials.

Instead of BLM with a planar arrangement of the membrane, spherical formations called *liposomes* or lipid vesicles are quite often used (see Fig. 93).

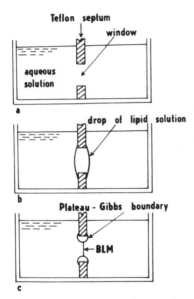

Fig. 92. Preparation of a bilayer lipid membrane. According to H. Ti Tien

Table 10   Thickness of BLM of different origin

Origin of BLM	Thickness (nm)	Method
Lecithin (egg)	5.6 ± 1.0	Capacitance
	6.2 ± 0.2	Light reflection
	6.9 ± 0.5	Electron microscopy
Oxidized cholesterol	4.0 ± 1.0	Light reflection
Glycerol diestearate	4.5 ± 0.5	Capacitance
	5.0 ± 0.5	Light reflection
Erytrocyte lipids	4.0 ± 1.3	Capacitance
Chloroplast lipids	10.5 ± 0.5	Light reflection
Brain lipids	6.0 ± 0.9	Electron microscopy

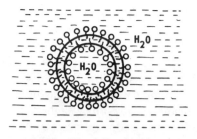

Fig. 93. A liposome

These tiny balls with a diameter of several tens of a nanometer consist of a bilayer film enclosing an aqueous solution. The liposomes are usually prepared by sonication (the effect of ultrasound) of a lipid suspension in water. Direct electrochemical measurement cannot be carried out on such tiny bodies. The membrane potential is measured only indirectly by means of a membrane soluble ion (tetraphenylphosphonium cation, for example). At equilibrium this ion is distributed between the bathing solution and the inside of the liposome according to the Nernst potential. Since the total concentration of this 'pilot' ion and its concentration in the bathing solution can easily be determined, the membrane potential is obtained from equation (3.1). A planar BLM cannot be investigated by molecular spectroscopy because of the small amount of substance in an individual BLM. This disadvantage is removed with liposomes because their suspensions can be quite concentrated. Thus, for application of ESR (= electron spin resonance, the spectra of which indicate the presence and properties of a substance with an unpaired electron), 'spin labelled' substances are embedded in the liposome membrane, such as a phospholipid with 2,2,6,6-tetramethylpiperidine-N-oxide (TEMPO)

bound to a long-chain alkyl group of the phospholipid. The properties of the spectrum signal of this substance in the environment of the membrane supply information on the movement and position of the spin-labelled lipid in the membrane.

Reference

1. H. Ti Tien, *Bilayer Lipid Membranes (BLM)*, Marcel Dekker, Inc., New York, 1974. D. Papahadjopoulos (ed.), 'Liposomes and their use in biology and medicine, *Annals of the New York Academy of Sciences*, Vol. 308, New York, 1978.

Passive Transport

Although the BLM is extremely thin, it exhibits enormous resistance at zero membrane potential between 18^8 and $10^{10}\,\Omega/cm^2$. The conductivity of such a simple 'unmodified' membrane results, perhaps, from the pores which are formed by the thermal motion of the molecules in the membrane. Water molecules enter these 'statistical' pores for a moment and enable the ions to pentrate through the membrane in a short time. When the membrane potential is increased from an internal voltage source the frequency and size of the pores grows and, finally, at a potential difference of several hundreds of millivolts the membrane breaks down.

The BLM conductivity increases when certain substances are added to the bathing solution or to the membrane alone. Most of these substances act on biological membranes in a similar way. The BLM is then an adequate model of a naturally occurring membrane. In this group belong, for example, hydrophobic ('lipophilic' = fat-liking) ions which alone are able to penetrate the membrane. Other substances cross the membrane through different mechanisms.

Figure 94 demonstrates several modes of electric charge transfer across the membrane. In addition to the simple transport of a hydrophobic ion, several transport mechanisms through ion-carriers (ionophores), channels, or pores are shown. These transport paths sometimes require an interaction between the carrier or the channel-forming molecules and the transported ion. This interaction often considerably depends on the individual properties of the ion, particularly on its size, and is, therefore, nearly always ion-selective. The rather narrow channel (formed as a rule by a molecule with helical shape) is not identical with a pore, which has such a large diameter that ions and solvent molecules can penetrate it regardless of their size.[1]

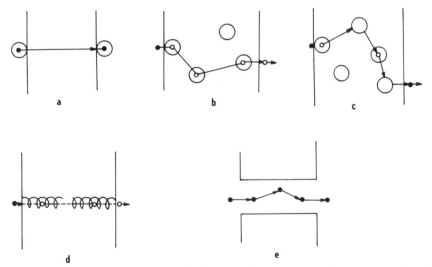

Fig. 94. Various mechanisms of passive ion transfer across a membrane. a, transfer of a lipophilic ion; b, simple carrier transport: the complex of the ion with the carrier migrate without a change across the membrane; c, carrier relay: the ion is exchanged between different carrier molecules when passing across the membrane; d, an ion-selective channel; e, a membrane pore which does not select between the transferred ions

Hydrophobic (lipophilic) ions which easily penetrate the membrane and considerably increase its conductivity include tetraphenylborate (see p. 145), picrate

and dipicrylaminate anions

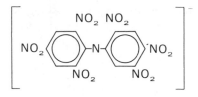

the long alkyl-chain tetraalkylammonium cations, etc.

For low concentrations, the membrane conductivity is directly proportional to the concentration of the lipophilic ion. When an electric potential difference is formed between the two bathing solutions, no rectilinear current-potential dependence is obtained. Such a dependence corresponding to Ohm's law would result in the membrane behaving as a simple 'passive' resistance. In fact, the actual current-potential dependence is similar to the diagram in Fig. 48. The laws governing the ion-transfer across the BLM and the electrode reaction rate are obviously analogous.

The carrier mechanism (usually a carrier relay—see Fig. 94) is connected with a group of substances discovered during the last thirty years.[2] The structure of these substances is usually cyclic. The groups stretching into the cavity of the cyclic (or *macrocyclic*) structure are polar, such as the carbonyl group, whereas the groups of the outer surface of the structure are non-polar, usually alkyl groups. The polar interior of the ionophore enables the cations to become complexed. In this way even complexes of alkali metal are formed, which otherwise avoid complex formation. On the other hand, the non-polar outer envelope helps to solubilize the complex in various organic low-polarity solvents where the free cations cannot be dissolved.

A typical substance in this readily growing family of natural and synthetic ionosphores is valinomycin (cf. p. 150) which is synthesized by the primitive fungus *Streptomyces fulvissimus*. Valinomycin is a cyclic depsipeptide with a 36-membered ring; its building units are alternatively aminoacids and α-hydroxyacids. Fig. 95 shows the bracelet structure of free valinomycin and the cylindrical structure of its potassium complex. The valinomycin network completely encircles the central cation. The complex with its lipophilic exterior can be dissolved in the membranes of cells and cell organelles and, in this way, can destroy an important metabolic function—oxidative phosphorylation (see p.

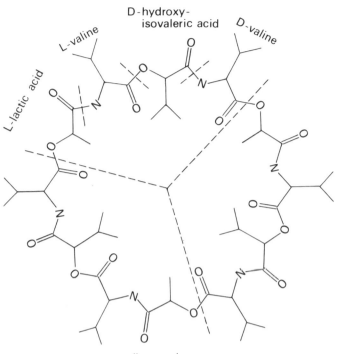

D-hydroxy-
isovaleric acid

L-valine

D-valine

L-lactic acid

valinomycin

179). This is the basis of *Streptomyces* antibiotic function. All processes that occur with the participation of valinomycin and related substances are ion-selective with respect to alkali metal cations, having different intensity in the presence of various ions. This ion-selectivity is linked to the complex stability, which is strikingly different for individual types of ions. When the alkali metal ion is small, such as the bare lithium ion, it easily fits into the internal cavity of valinomycin. On the other hand, a great deal of energy is needed to liberate this ion from its large hydration sheath, resulting in the negligible stability of the valinomycin complex of lithium. The potassium ion has a larger ion radius than the lithium ion but still fits into the cavity, so that the solvation is much weaker than for lithium and, therefore, the tendency to complex formation is much larger. The caesium ion has virtually no hydration sheath but the radius of the free ion is larger than that of the cavity, so that the macrocyclic structure is strained and the binding in the complex is less advantageous than with potassium. Consequently, the caesium complex of valinomycin is less stable than the potassium complex. A completely analogous dependence on the ion radius, as demonstrated by the stabilities of complexes, is exhibited by the resistances of BLM (see Fig. 96) and by the membrane potentials in the presence of valinomycin and the salts of different alkali metal cations.

Above 30–40°C, the BLM behaves like a liquid with oriented molecules. This state of aggregation is termed *liquid-crystalline*. When the temperature falls below a critical value characteristic for any BLM material, the aggregation of the

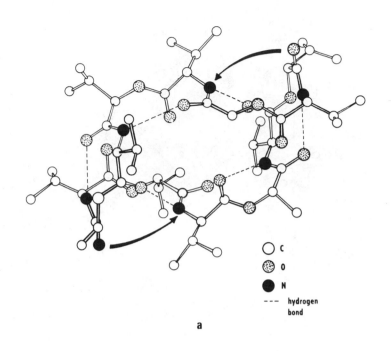

O C

O O

● N

- - - hydrogen
bond

a

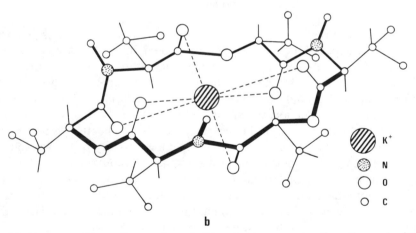

K⁺

N

O

C

b

Fig. 95. Structure of free valinomycin and of its potassium complex (from the work of V. T. Ivanov and Yu. A. Ovchinnikov)

membrane becomes crystalline (gelation). Then the motion of both the molecules that form the membrane or that are dissolved in it is considerably restricted. In the presence of valinomycin or other ionophores the conductivity of the membrane strongly decreases as the ionophore loses its function. On the contrary, a membrane that contains the antibiotic valin-gramicidin A, retains its original conductivity value even after a temperature decrease below the critical

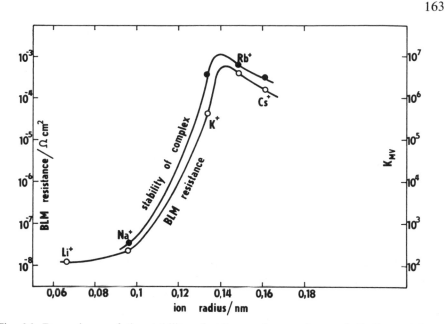

Fig. 96. Dependence of the stability of valinomycin complexes of alkali metal ions (according to A. Hofmanová, J. Koryta, and M. Březina) and BLM resistance in solutions containing valinomycin and alkali metal chlorides (according to P. Läuger) on the cation radius

value. Gramicidin A is a peptide with 15 building units and with a helical structure

HC=0
|
L-Val -Gly—L-Ala—D-Leu—L-Ala—D-Val—L-Val—D-Val
                                              |
NH—L-Try—D-Leu—L-Try—D-Leu— L-Try—D-Leu—L-Try
|
(CH$_2$)$_2$
|
OH

(Val = valine, Gly = glycine, Ala = alanine, Leu = leucine, Try = tryptophane)

The helix has a length such that two molecules of the peptide attached together at their ends just extend beyond the width of the BLM (Fig. 94D). Thus, these two helices form a transmembrane channel through which the ions penetrate across the membrane. This channel allows the passage of ions even if the neighbouring lipid molecules are in a crystalline state. The outer surface of the channel must be lipophilic so that it can be placed inside the membrane. Its structure must be sufficiently mobile to enable the shape of the turns of the helix to adjust to the permeating ions. The binding of the ions in the channel must be weaker than for

164

the ionophore because the ion must be released quickly so that it can pass through the channel. The radius of the turns decides which ions will cross the membrane easily. The selectivity of the gramicidin channel for various monovalent ions obeys the sequence

$$H^+ > NH_4^+ \geq Cs^+ \geq Rb^+ > K^+ > Na^+ > Li^+$$

When a small amount of gramicidin A is dissolved in the BLM (the substance is insoluble in water) and the membrane conductivity is measured with a sensitive fast-recording instrument, the time-dependence shown in Fig. 97 is obtained. The conductivity fluctuates stepwise so that the height of the individual steps is approximately the same. One step corresponds to one channel in the BLM which is open for a definite time (the mechanism of the opening and of the closing is not known) and enables the ions to pass across the membrane under the influence of the electric field.

The interior of the helix of some channel-forming peptides is partitioned by hydrogen bridges between the CO and NH groups belonging to different aminoacid units so that the membrane channel is impermeable. Under the influence of an electric field the dipoles of the carbonyl groups align along the direction of the field and the hydrogen bridges are disrupted, so that the ions can penetrate through the membrane. Therefore the membrane conductivity increases considerably when the electric field strength increases.[3] A similar property is also exhibited by the nerve cell membrane (see p. 172) where there are probably similar transmembrane channels.

Polyene-type antibiotics, like filipin, nystatin, or amphotericin B and some other substances, form pores in the membrane. The conductivity of the membrane strongly depends on the concentration of the antbiotic (dependences of $G \sim c^5$ and even $G \sim c^{10}$ have been observed). Thus, the pores in the membrane are formed as a group of several antibiotic molecules. These pores exhibit low selectivity because various cations, water, and other non-electrolytes can permeate indiscriminately. However, some selectivity towards anions was observed.

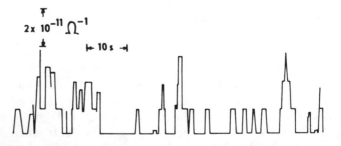

Fig. 97. Time dependence of the BLM conductivity in the presence of gramicidin A. The individual steps correspond to opened transmembrane channels which close after some period of time. According to D. A. Haydon

The transport of matter across the membrane has so far depended on only two 'driving forces': the electric field strength and the concentration gradient of the substance transported across the membrane. These two forces can be combined in the gradient of the electrochemical potential of the species concerned (cf. p. 55). This is a characteristic property of the *passive transport* which takes place in the direction of the drop (negative gradient) of the electrochemical potential. In biological membranes this sort of transport is far from the only transport phenomenon. Very often, transport takes place in a direction opposite to the drop of the electrochemical potential ('up-hill transport') or takes place without any electrochemical potential gradient, etc. This sort of transport is termed *active transport*.[4]

References

1. D. A. Haydon and B. S. Hladky, *Quart. Rev. Biophys.*, **5** (1972) 187.
2. Yu. A. Ovchinnikov, V. I. Ivanov, and M. M. Shkrob, *Membrane Active Complexones*, Elsevier, Amsterdam, 1974.
3. D. W. Urry, *Proc. Nat. Acad. Sci. USA*, **69** (1972) 1610.
4. For a review see A. Kotyk and K. Janáček, *Cell Membrane Transport: Principles and Techniques*, 2nd edn, Plenum, New York, 1975.

Active Transport

The main ionic component in the organism of a frog, sodium ions, are not washed out of his body when the frog stays in water. This phenomenon depends on the active transport of sodium ions in the skin in the direction from the epithel to the corium, i.e. to the inside of the skin.

In an experiment[1] the skin of the common frog, *Rana temporaria* representing a membrane is fixed between two compartments of a cell containing identical electrolyte solutions (usually 0.1 M NaCl with additions of KCl, $CaCl_2$, and $NaHCO_3$). In the absence of electric current the electric potential difference between the electrolyte in contact with the outer side of the skin and the other electrolyte is -50 mV. Obviously, an electric potential difference is formed, although the composition of both the solutions is the same. This potential difference can be decreased to zero when an electric current flows in the direction from the outer side to the inner side of the membrane. When the solution on the outer side of the skin is labelled with the radioactive isotope ^{22}Na, it is found that sodium ions are transported from the outer side to the inner side of the skin.

Sodium transport occurs in this direction even if the compartment in contact with the inner side of the skin contains a higher sodium ion concentration than the other compartment.

When the temperature of the solution in the cell is increased, the current increases far more than would correspond to mere diffusion or ion migration. When substances inhibiting metabolic processes, e.g. cyanide or ouabain, are added to the cell solution, the current decreases. For example, in the presence of 10^{-4} M ouabain (a digitalis glycoside)

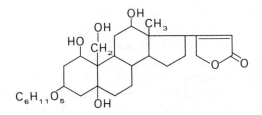

it attains only 5% of the value in the absence of ouabain. For the sodium transport in the absence of a concentration gradient or even against the concentration drop, there is no source of energy in the system other than the chemical reaction occurring in the skin. The decisive role of the chemical reaction for Na$^+$ transport is also demonstrated by the temperature dependence of the transport rate. As a rule, a chemical reaction has much a larger activation energy than diffusion or migration transport. The activation energy of these processes is determined by the activation energy for particle mobility which amounts about 13 kJ/mol, resulting, for example, in a 2.5% increase in the diffusion coefficient when the temperature is increased from 25 to 26°C. This is much less than the usual increase in the chemical reaction rate in the same temperature interval.

Thus, active transport takes place as a result of a decrease in the Gibbs energy of a chemical reaction, usually a metabolic reaction connected with oxidation of a certain substrate (a biochemical 'fuel') or with a rupture of an energy-rich ('macroergic') polyphosphate bond in adenosine triphosphate (ATP) with the products adenosine diphosphate (ADP) and phosphate ion

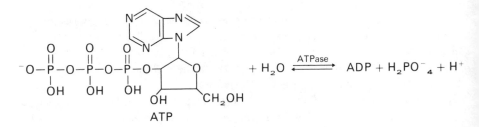

The hydrolysis of adenosine triphosphate is catalysed by enzymes termed ATPases (some of them are reversible and even catalyse ATP synthesis of ADP and phosphate ions). To function, ATPases require the presence of alkali or alkaline earth metal ions.

In some cases the source of energy for active transport is the concentration gradient of a substance other than that actively transported (for example, for the active transport of glucose across the cell membrane of the small intestine, the necessary energy is supplied from the concentration gradient of sodium ions). Another term for active transport is the 'pump'.

Let us return to Na^+ transport in the frog skin. This process consists of several steps. The epithel of the frog skin is made up of three layers of cells. The sides facing the outer solution are permeable for sodium ions. In the inner side of these cells are regions with built-in molecules of Na^+, K^+-ATPase which is activated by sodium and potassium ions in the presence of magnesium ions. This enzyme consists of two components of which α has a molar weight larger than 100,000 and β about 50,000. This is a genuine site where the sodium ions are removed from the epithel cells and substituted with potassium ions, which simultaneously preserve the electroneutrality of the solutions because the sodium and potassium fluxes are opposite in direction and equal in absolute value. This exchange requires energy, of course, which is supplied by ATP hydrolysis (one ATP hydrolysed molecule corresponds to three transported sodium ions). The potassium ions whose intracellular concentration increases compared with the intercellular liquid, leave the cells by passive transport (see Fig. 98). On the contrary, as a result of the exchange reaction mediated by the ATPase, the sodium ion concentration in the epithel is lower than in the ambient solution, and therefore the first step for sodium transport is its passive transport from the outer solution into the cell.[2]

While the mechanism of passive transport is quite often clear (see p. 158), no detailed mechanism for the coupling of transport of matter and energy supply by a chemical reaction is known. The function of Na^+, K^+-ATPase is possibly linked to temporary phosphorylation of the protein by phosphate ions transferred from ATP. The energy gained enables the enzyme to undergo a conformational change which is connected with sodium ion transport from the inside of the cell into the intercellular liquid and with potassium ion transport in the opposite direction (Fig. 99).

References

1. J. Bureš, M. Petráň, and J. Zachar, *Electrophysiological Methods in Biological Research*, 3rd edn, Academic Press, New York, 1967.
2. S. L. Bonting, 'Sodium–potassium activated adenosinetriphosphatase and cation transport', in: *Membranes and Ion Transport* (E. E. Bittar, ed.), Wiley–Interscience, New York, 1970, chapter III.8.

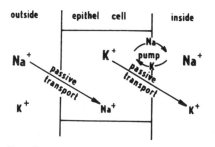

Fig. 98. A scheme of sodium ion transport across frog skin. Upper case letters indicate higher ion concentration

168

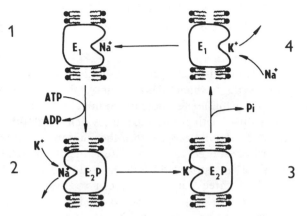

Fig. 99. Hypothetical function of Na^+, K^+-ATPase. Position 1: the enzyme binds Na^+. Position 2: after phoshorylation it undergoes a conformational change and transfers Na^+ on the other side of the membrane. Position 3: the enzyme binds K^+. Position 4: after release of the phosphate anion the enzyme acquires the original conformation and transfers K^+ to the other side of the membrane

Electrochemistry of Nervous Activity

Sodium and potassium are elements of paramount importance for a number of processes in organisms. Whereas sodium dominates over potassium in the organism as a whole, inside the cells the potassium concentration is usually larger than that of sodium. For example, in the frog muscle the intercellular liquid contains 120 mM Na^+ and 2.5 mM K^+, while the intracellular concentrations are 9.2 mM Na^+ and 140 mM K^+.

These two ionic species play a decisive role in the activity of the nervous system.[1] Information transfer in the nerves is based on the transmission of uniform signals, called action potentials, along the nerve fibre. The nerve cell has a body with star-like projections which comprises the cell nucleus, and a long tail termed an axon (Fig. 100). In some parts the axon is wrapped in a multiple myelin layer so that its membrane only contacts the intercellular liquid in the nodes of Ranvier. The nerve impulses are transferred between nerve fibres in the synapses. An important experimental material is the axons of certain cephalopodes, like squid or of another sea mollusc Aplysia, which are rather thick (diameter as large as 1 mm) and are not covered by a myelin layer.

The inside of the axon considerably differs from the intercellular liquid. Thus, for example, the squid axon contains 0.05 M Na^+, 0.4 M K^+, 0.04–0.1 M Cl^-, 0.27 M isethionate, and 0.075 M aspartate, while the intercellular liquid comprises 0.46 M Na^+, 0.01 M K^+, and 0.054 M Cl^-. When the nerve cell is at rest, a rest membrane potential is established at the axon membrane, $\Delta\varphi_M = \varphi(2) - \varphi(1)$, where $\varphi(2)$ denotes the internal potential of the axon while $\varphi(1)$ is the potential of the intercellular liquid (sometimes $\varphi(2)$ is denoted φ_i and $\varphi(1)$, φ_o). The value of the rest potential is about -70 mV.

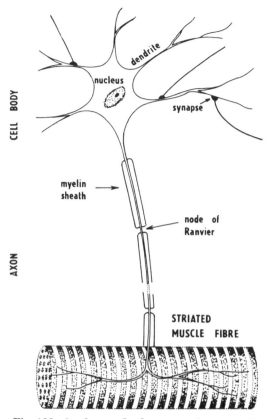

Fig. 100. A scheme of a frog motoric nerve cell

A theory of the rest potential of the axon membrane was developed by Goldman, Hodgkin, Huxley, and Katz based on the assumption that the electric field strength in a thin membrane is constant and that the ion transport in the membrane can be described by the Nernst–Planck equation. However, this approach does not seem realistic since the ions are transmitted across the membrane through channels that are specific for individual ionic types. In transport through a channel with molecular dimensions it is hardly possible to speak about diffusion because the ions jump across the membrane, which requires overcoming an energy barrier.

Ion jumps will be described by the same equation as the electrode reaction on p. 92. Assume that the rest membrane potential is determined solely by transfer of potassium, sodium, and chloride ions. For the individual fluxes of the ions it then follows that ($\alpha = \frac{1}{2}$)

$$J_{K^+} = k_{K^+}^0 \, c_{K^+}(1) \exp\left[-F\,\Delta\varphi_M/2RT\right)\right] - k_{K^+}^0 \, c_{K^+}(2) \exp\left[F\,\Delta\varphi_M/(2RT)\right]$$

$$J_{Na^+} = k_{Na^+}^0 \, c_{Na^+}(1) \exp\left[-F\,\Delta\varphi_M/(2RT)\right] - k_{Na^+}^0 \, c_{Na^+}(2) \exp\left[F\,\Delta\varphi_M/(2RT)\right]$$

$$J_{Cl^-} = k_{Cl^-}^0 \, c_{Cl^-}(1) \exp\left[F\,\Delta\varphi_M/(2RT)\right] - k_{Cl^-}^0 \, c_{Cl^-}(2) \exp\left[-F\,\Delta\varphi_M/(2RT)\right]$$

Since, at rest, no electric current flows across the membrane, the flux of chloride ions compensates that of sodium and potassium ions

$$J_{Na^+} + J_{K^+} = J_{Cl^-} \tag{3.24}$$

Combining equations (3.23) and (3.24) yields the expression for the membrane potential

$$\Delta\varphi_M = (RT/F)\ln\frac{k_{K^+}^0 c_{K^+}(1) + k_{Na^+}^0 c_{Na^+}(1) + k_{Cl^-}^0 c_{Cl^-}(2)}{k_{K^+}^0 c_{K^+}(2) + k_{Na^+}^0 c_{Na^+}(2) + k_{Cl^-}^0 c_{Cl^-}(1)} \tag{3.25}$$

Ion transfer is characterized by standard reaction rate constants $k_{K^+}^0$, $k_{Na^+}^0$, and $k_{Cl^-}^0$, which can be identified with the permeabilities for these ions. This simple approach leads to the same result as the treatment of the problem by Hodgkin, Huxley, and Katz. Equation (3.25) satisfactorily explains the value of the rest membrane potential found in experiments, assuming that the permeability of K^+ is larger than that of Na^+ and Cl^-, so that the deviation from the Nernst potential for potassium ions is not very large. On the other hand, the permeabilities for the other ions are not negligible. Consequently, the axon at rest would lose potassium ions and the intracellular sodium concentration would correspondingly grow. This, of course, does not happen because the Na^+, K^+-ATPase is again active at the expense of ATP hydrolysis, the enzyme transfers the potassium ions from the intercellular liquid into the axon and the sodium ions in the opposite direction. Since this process is not connected with current flow nor does it influence the membrane potential, it is termed the *electroneutral pump*. In addition, active transport can also occur without 'ion for ion' exchange so that a change in the membrane potential results. This *electrogenic pump* is observed, for example, with muscle fibres when they are kept in a potassium-free, sodium-rich medium for a certain time. They then become loaded with sodium ions through exchange of intercellular potassium for extracellular sodium ions and, after returning to a medium whose composition corresponds to a common intercellular liquid, sodium is extruded from the cells by active transport to a degree such that the membrane potential shifts to more negative values (the cell membranes hyperpolarized). The hyperpolarization is removed by ouabain (cf. p. 166).[2]

The basic experiments with electric stimulation of nerve cells were carried out by Cole, Hodgkin, and Huxley who worked with giant squid axons. Their experimental arrangement is shown in Fig. 101. The membrane potential is measured by two reference electrodes, e.g. AgCl/Ag, which are attached to the liquids under investigation by micropipettes filled with saline (0.9 % NaCl) and gelified with agar–agar (cf. p. 76). One of the micropipettes is immersed in the intercellular liquid close to the axon surface while the other is impaled into the membrane so that its orifice is placed inside the axon in the intercellular liquid. The auxiliary electrodes are placed either in the positions shown in Fig. 101 or one of them is slid into the axon from the side (after cutting off the end of the axon).

As already mentioned, the rest membrane potential is about $-70\,mV$. When the cell is excited with small rectangular current pulses the membrane potential

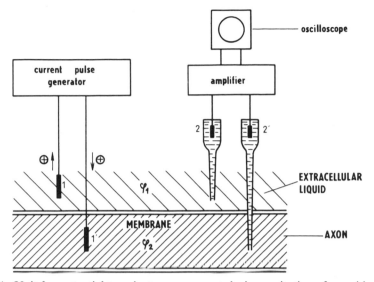

Fig. 101. Unit for potential transient measurement during excitation of a squid axon by current pulses from electrodes 1 and 1'. 2 and 2' are micropipettes. According to A. L. Hodgkin and A. F. Huxley

changes, as shown in Fig. 102 (the current flowing into the cell is plotted in the diagram). The extent of this change depends on the amount of electricity transferred in the current impulse. With negative current pulses the membrane potential shifts to more negative values—it is hyperpolarized. The current flowing in the opposite direction (a positive current) has a depolarizing effect. The potential drops to zero and then increases to positive values. When the pulse exceeds a threshold value a steep potential increase appears. The peak formed is termed the spike or action potential. Its height does not depend on a further increase of the current pulse. Sufficiently large excitation of the membrane results in a large increase in the membrane permeability for sodium ions so that, finally,

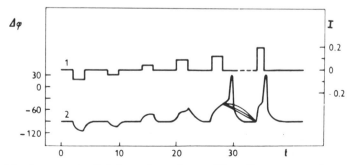

Fig. 102. Membrane potential dependence (curve 2) on time course of current pulses (curve 1). The spike sets in at $t = 30$ s. According to B. Katz

the membrane potential almost acquires the value of the Nernst potential for sodium ions ($\Delta\varphi_M = +50\,\text{mV}$). A potential drop to the rest value is accompanied by a temporary influx of sodium ions from the intercellular liquid into the axon.

Basic information on the nature of the currents flowing across the membrane was acquired with the four-electrode potentiostatic method (see p. 146 and Fig. 82). In electrophysiology (where it was first introduced) it is termed the 'voltage clamp' method. In this approach the membrane potential is kept constant and the current flowing across the membrane is measured. In this way important information on sodium, potassium, calcium, and other ions can be obtained. There is now no doubt about the presence of specific channels through which different sorts of ions are transported, while little is known about the structure of these channels. It appears that they may be peptidic spirals lying across the membrane (cf. p. 163) with specific groups at their orifices that permit the opening and closing of the channels (gates).

The function of the gates depends on the membrane potential which explains, for example, the avalanche-like increase of sodium permeability at the spikes. The sodium channel, or, preferably, the sodium gate, is blocked by tetrodotoxin, a poison present in some tropical fish and in newt species

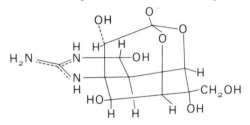

Its physiological action is manifested by breath paralysis. The potassium channel is blocked by triethylnonylammonium ion and its homologues. The calcium channel is inhibited by cadmium and the gate is destroyed by fluoride. Magnesium ions can also use the calcium channel for crossing the membrane.[3]

In Fig. 103 an example of the time-dependence of the membrane current is shown for the case when the potassium channel is blocked. The imposed membrane potential was changed by a rectangular voltage pulse from -78 to $-18\,\text{mV}$. Obviously, first the gates of the sodium channel gradually open but they then remain in the open position for different time intervals so that the channels are inactivated after some time.

The energy for nervous activity is stored in the form of unequally distributed sodium and potassium ions in the intercellular and intracellular space. This energy can be liberated by small amounts of energy needed for opening the gates. That the energy of the gradients of alkali metal ions can be quite large is demonstrated by the example of the 'mysterious' phenomena of electric fish (cf. p. 125). For centuries naturalists have been interested in these fish: the electric eel (see Fig. 104), the torpedo, and others.* Comparatively strong discharges (with

---

* M. Faraday used to keep a living torpedo in his laboratory of the Royal Institution for experiments aimed at demonstrating the unity between 'biological' and other sorts of electricity.

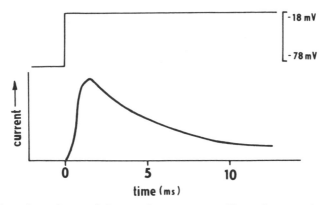

Fig. 103. Time dependence of the membrane current. Since the potassium channel is blocked the current corresponds to sodium transport. The upper line represents the time course of the imposed potential difference

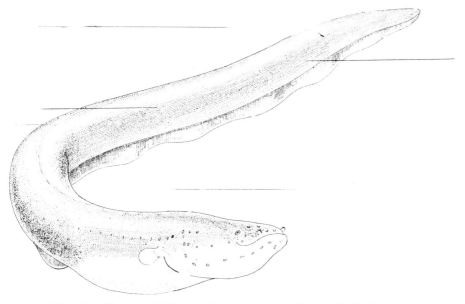

Fig. 104. Electric eel (*Electrophorus electricus*). Drawn by E. Opatrný

the electric eel the electric potential difference between the head and the tail in the air is as much as 600 V) originate from the electric organ which occupies about four-fifths of the body of this fish. The electric organ consists of several thousands of platelet cells (see Fig. 105) which contain potassium and sodium ions in similar concentrations as in the nerve cells. Only the flat upper sides of the platelets are innervated. When the fish is not excited a rest potential builds up around the whole membrane and the electric organ does not function as a voltage source. On excitation a spike sets in on the upper sides of the membranes while the lower

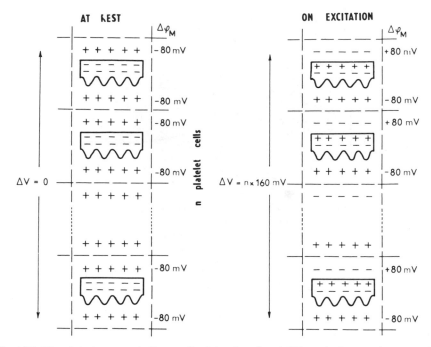

Fig. 105. The electric organ (a 'battery') of the electric eel. When the battery is at rest the platelet cell membranes all have the rest potential (left-hand side). On excitation a spike sets in on the folded parts of the membranes

folded sides remain at rest. The electric organ then becomes a battery of galvanic cells (with an EMF of about 150 mV each) connected in series. The fish can use the large resulting electric potential difference for stunning or even killing other fish[5].

References

1. B. Katz, *Nerve, Muscle and Synapse*, McGraw-Hill, New York, 1966.
2. G. S. Taylor, D. M. Paton, and E. E. Daniel, *J. Gen. Physiol.*, **56** (1970) 3.
3. W. Ulbricht, *Biophys. Structure Mech.*, **1** (1974) 1.
   P. G. Kostyuk and O. A. Krishtal, 'Ionic and gating currents in the membranes of nerve cells', in: *Membrane Transport Processes*, Vol. 2 (D. C. Tosteson, Yu. A. Ovchinnikov, and R. Latorre, eds.), Raven Press, New York, 1978.
5. M. V. L. Bennett, *Ann. Rev. Physiol.*, **32** (1970) 471.

Muscle Activity

Locomotion in higher organisms as well as other mechanical activities is made possible by striated skeleton muscles. The basic structural unit of a muscle is the muscle cell, the muscle fibre (see Fig. 106). The muscle fibre is enclosed by the sarcoplasmatic membrane or sarcolemma (Greek *sarx* = flesh, *lemma* = substance). This membrane invaginates into the interior of the fibre by

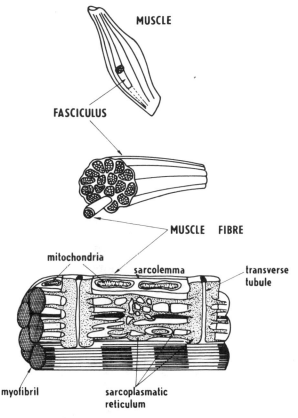

MUSCLE

FASCICULUS

MUSCLE FIBRE

mitochondria

sarcolemma

transverse tubule

myofibril

sarcoplasmatic reticulum

Fig. 106. Striated muscle structure

transversal tubules which are filled with the intercellular liquid. The inside of the fibre consists of the actual cytoplasm (sacroplasm) with inserted mitochondria, sacroplasmatic reticulum (network) and myofibrils (Greek *myon* = muscle). The myofibrils are the organs of muscle contraction and extension. The sarcoplasmatic reticulum is a membrane system enveloping the myofibrils and leading into the neighbourhood of the tubules through the cisterns.

The genuine excitation of the muscle fibre is contingent on the transfer of sodium and potassium ions across the sarcolemma. The origin of the spike is the same as in the axon, but subsequently the basic role is played by calcium ions. The depolarization of the sarcolemma is accompanied by a decrease in the potential between the transversal tubules and the adjacent parts of the sarcoplasma resulting in changes in the membrane potential of the sarcoplasmatic reticulum in the neighbourhood. The calcium concentration in the sarcoplasm at rest is lower than 10^{-7} mol/dm^3 while after excitation it increases to a value of about 10^{-4} mol/dm^3. This sudden increase in the calcium concentration activates the myofibrils and elicits their contraction. The myofibrils are composed of parallel thin filaments consisting of the protein actin and of thick filaments built of another protein, myosin. The relative movement of these filaments, on which muscle contraction is based, depends on energy supply by hydrolysis of ATP. This movement is hindered by a further protein, troponin, which lies between the thick and thin filaments. The reaction between calcium ions and troponin changes its conformation resulting in the disappearance of the inhibition and firing of the contraction mechanism. When the contraction stops the sarcoplasmatic calcium is removed into the sarcoplasmatic reticulum by active transport as long as its concentration in the sarcoplasm drops to the original value.

Bioenergetics

As already mentioned on p. 125, all the energy necessary for life comes from the sun—directly or indirectly. The basic energy processes in all living things, or bioenergetics, are photosynthesis of carbonaceous fuels, sugars, and conversion of these fuels into more serviceable material, ATP (see Fig. 71).

The latter step, cell respiration, is relatively simpler and, therefore, will be dealt with first. Here, the Gibbs energy gained by oxidation of carbonaceous substances in cell breathing is accumulated in the reaction of adenosine diphosphate (ADP) with the phosphate ion or, more exactly, the dihydrogen phosphate ion $H_2PO_4^-$, as the pH of the medium is around 7, to give adenosine triphosphate (ATP—see p. 166). This process takes place in membranes of mitochondria, organelles present in cells of the vast majority of multicellular organisms (see Figs. 68 and 69). The mitochondrion has two membranes. The internal membrane invaginates into the inner space of the organelle (matrix space) through crests (Latin: *cristae*). The matrix space contains a considerable amount of various enzymes while the space between the internal and the external membrane of the mitochondrion (intracristal space) comprises mainly low-molecular substances including adenine nucleotides (ATP, ADP, and AMP, adenosine monophosphate). The oxidation of a carbonaceous substrate by oxygen cannot occur directly because the energy set free would be converted into heat. The process must be carried out in an almost reversible way, with the help of a number of oxidation–reduction systems of low solubility in water which form the electron transport chain. The components of the electron transport chain are built into the internal membrane of the mitochondria. They include the nicotinamide adenine dinucleotide system ($NAD^+/NADH$), several flavoprotein systems, particularly FMN, the flavin mononucleotide, the iron–sulphur proteins (Fe ions are bound in complexes with cysteine of the protein and inorganic sulphide), coenzyme Q (ubiquinone), and a number of cytochromes (for cytochrome c, see p. 77 and Fig. 40). These redox systems are combined, sometimes with phospholipids and other molecules, into several blocks and ordered across the membrane according to their oxidation–reduction potentials. (The oxidation–reduction potentials of these systems have been measured, if at all, in solution so that their actual value in the membrane will differ considerably.) The ordering of redox components enables the oxidation of the carbonaceous substrate to occur 'vectorially',[1] which means that the chemical change is spatially oriented (see Fig. 107).[2]

The oxidation of the substrate by oxygen comprises an array of partial oxidation steps, which Mitchell termed loops. Each loop consists of two basic processes, one oriented from the inner membrane surface facing the matrix space to the surface contacting the intracristal space, and the other in the opposite direction (Fig. 107). In the former process, electrons are transferred together with protons which are extruded to the intracristal space, while in the latter process only electrons are transported. The first step of the first loop is the reduction of NAD^+ by the carbonaceous substrate SH_2, resulting in the joint transfer of H^+ and electrons. The second step of the last loop is the reduction of oxygen by the cytochrome a_3 system containing copper. Thus, the protons coming from the matrix space are accumulated in the intracristal space which is connected with a decrease in the pH and an increase in the membrane potential ($\Delta\varphi_M$). As shown by Mitchell, the major part of the Gibbs energy change connected with oxidation of the substrate by oxygen is thus stored in the form of increased electrochemical potential of the protons in the intracristal space

$$\Delta\tilde{\mu}_{H^+} = F\Delta\Delta\varphi_M - 2.3RT\Delta pH,$$

where $\Delta\Delta\varphi_M$ is the increment of the membrane potential.

How is this accumulated energy utilized? The inner mitochondrial membrane contains a protein, H^+-ATPase, which is able to synthesize ATP using a proton gradient across the membrane. Fig. 108 shows a scheme of the structure of this enzyme. A hydrophobic section F_0, which is built into the hydrophobic part of the bilayer, is responsible for the proton transport from the intracristal space.

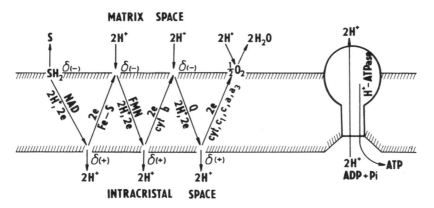

Fig. 107. A simplified scheme of electron transfer from the substrate to oxygen via electron transfer chain enzymes of the inner mitochondrial membrane coupled with hydrogen ion accumulation in the intracristal space. SH_2 denotes a carbonaceous substrate, for example succinic acid; NAD, nicotinamide adenine dinucleotide; Fe–S, the iron–sulphur protein; FMN, flavin mononucleotide; Q, ubiquinone, cyt cytochrome; and Pi, the inorganic phosphate ion. According to P. Mitchell

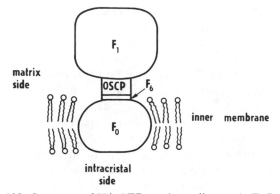

Fig. 108. Structure of H^+-ATPase. According to A. E. Senior

This component is linked by protein F_6 and OSCP (oligomycin-sensitivity-conferring protein) to the ATP synthesizing section F_1. The ATP synthesizing function of the enzyme is inhibited by the antibiotic, oligomycin. H^+-ATPase acts in the reversible way, i.e. under reversed proton gradient or at low phosphate concentration it converts ATP to ADP and the phosphate ion. The details of the molecular mechanism of the synthesis and hydrolysis of ATP are being intensively investigated at present. Obviously, intermediate phosphorylation of the enzyme alone plays an important role here.

In the presence of various substances the oxidative phosphorylation is uncoupled, i.e. while the carbonaceous substrate is oxidized no ATP is synthesized but is hydrolysed to ADP and phosphate. The uncouplers of oxidative phosphorylation belong to two groups of ionophores, one transferring protons, the other alkali metal ions. The first group includes the acid anions listed in Table 11. Because of spreading of the negative charge around the whole structure, the anion can penetrate the lipidic part of the mitochondrial membrane in the direction of increasing electric potential. When it reaches the intracristal side with a high hydrogen ion concentration, it is transformed to a non-dissociated form of the acid which is then transported in the opposite direction to the matrix space. Here it splits-off the proton which has been acquired in the intracristal space.

Both the transport of the anion into the intracristal space and of the acid back are passive and therefore decrease both the membrane potential and the hydrogen activity in the intracristal space. In this way the energy gained in the electron-transfer chain is dissipated as heat. In consequence of the decrease in the H^+-activity in the intracristal space, the action of the ATPase is reversed and ATP is hydrolysed.

The second group substances are the known macrocyclic complex-formers like valinomycin. With this species the most pronounced uncoupling occurs in the presence of potassium ions (cf. its effect on the potential of the ion-selective electrode, p. 150, and on the membrane conductivity, p. 161). As a result of its lipophilicity the electrically uncharged ionophore freely diffuses in the inner

Table 11 'Classical' uncouplers of oxidative phosphorylation

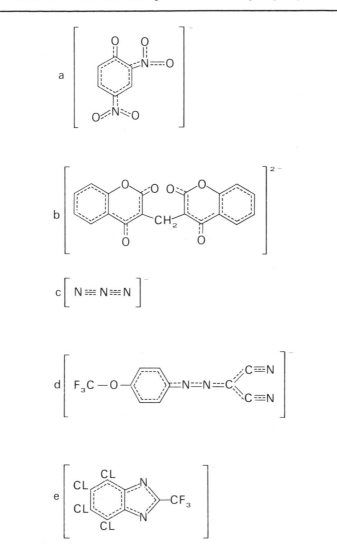

a, 2,4-dinitrophenolate; b, dicoumarate; c, azide; d, carbonyl cyanide p-trifluoro-methoxyphenylhydrazone; e, tetrachloro-2-trifluoromethylbenzimidazole. According to P. Mitchell.

membrane of the mitochondrion. On the intracristal side of the membrane it combines with the potassium ion. The complex is then transported passively to the matrix space. This process, which in fact simulates the proton transport, decreases the membrane potential. The elucidation of the uncoupling of oxidative phosphorylation represents a notable support for the Mitchell theory.

Cell respiration is a basic step through which energy is supplied to the organism, but an absolutely necessary source for this oxidation process is the 'fuel', i.e. the carbonaceous compounds, primarily sugars, forced in photosynthesis, the energy of sunlight being the primary energy source for all living organisms. In green plants the light energy is supplied to an oxidation–reduction reaction between sufficiently abundant components, water and carbon dioxide. This reaction occurs according to the equation

$$6\,CO_2 + 6\,H_2O \xrightarrow[\text{chlorophyll}]{hv} C_6H_{12}O_6 + 6\,O_2$$

while in photosynthetic bacteria the reaction

$$6\,CO_2 + 12\,H_2S \xrightarrow[\text{bacteriochlorophyll}]{hv} C_6H_{12}O_6 + 12\,S + 6\,H_2O$$

takes place. The latter reaction cannot, of course, become a universal photosynthetic process because hydrogen sulphide is unstable, is not present in sufficient amounts, and is produced mainly by biological processes. The only substance which can be exploited for carbon dioxide reduction, and is present in the biosphere independently of biological processes, is water. The most important catalysts of photosynthesis are various sorts of porphyrin complexes of magnesium, the chlorophylls. The halophilic (Greek *hals* = salt) purple microorganism, *Halobacterium halobium*, contains the protein, bacteriorhodopsin (Greek *rhodos* = purple, *opsis* = sight), which is structurally related to the visual pigment, rhodopsin, as a photosynthetic substance.[4]

Photosynthesis in green plants proceeds through two basic processes. The dark process (the Calvin cycle) is the reduction of carbon dioxide by a strong reductant, nicotinamide adenine dinucleotidephosphate, with a supply of energy gained by the transformation of ATP to ADP

$$6\,CO_2 + 12\,NADPH_2 + 6\,H_2O \longrightarrow C_6H_{12}O_6 + 12\,NADP$$
$$ATP \nearrow \qquad \searrow ADP + P_{inorg}$$

The light process is a typical membrane process. The membrane of thylakoids, which are organelles forming a larger organelle, the chloroplast (see Fig. 70), contain chlorophyll (or, more exactly, an integrated system of several chlorophyll-type pigments and carotenes) and an enzymatic system which make possible the accumulation of the energy of sunlight in the form of nicotinamide adenine dinucleotidephosphate and ATP (this kind of formation of ATP from ADP and a phosphate ion is termed photophosphorylation).[5] The photosynthetic apparatus of blue-green algae is embedded in their cell membranes.

As shown in Fig. 109, the light process occurs in two photosystems. On absorption of a light quantum by the chlorophyll of System II, charge separation sets in, i.e. an electron–hole pair is formed. The hole is annihilated by an electron from the enzyme containing manganese in its prosthetic group. The extraction of a total of four electrons (connected with absorption of four light quanta by

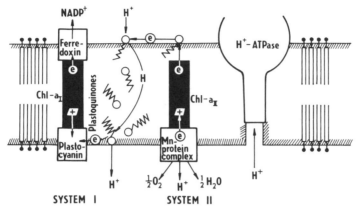

Fig. 109. A simplified scheme of the light process of photosynthesis. According to H. T. Witt

System I) is necessary for the evolution of the oxygen molecule. This process requires the concerted action of four manganese-containing sites in the enzyme (each containing a pair of Mn atoms) where divalent manganese is oxidized to the trivalent form and hydroxide ions bound in complexes with manganese are transformed to bound superoxide or hydroperoxide groups from which, in the first step, molecular oxygen is set free.[6] At the same time protons cross the inner side of the thylakoid membrane and accumulate inside the thylakoid.

The energy of the electron formed in System II is not high enough to reduce NADP so that another light quantum is required. To accomplish this the electron reacts first with hydrophobic quinone, the plastoquinone, adsorbed on the outer surface of the thylakoid membrane. The reduced form, the plastohydroquinone, accepts two protons from the outer solution. Both the electrons and protons are transferred to further plastoquinone molecules and the electrons finally arrive at the enzyme, plastocyanine, from where they annihilate the holes formed in System I after absorption of light. At the same time the protons originating in the outer solution are transferred into the thylakoid. The high energy electron set free in System I reduces the Fe–S protein, ferredoxin, and through a chain of several other enzymes attacks the oxidized nicotinamide adenine dinucleotidephosphate. Part of the energy gained is stored in the form of increased electrochemical potential of the hydrogen ions in the interior of the thylakoid. This energy is then supplied to the H^+-ATPase. In contrast to cell respiration, where the energy from the oxidation of a carbonaceous substrate is accumulated in a single 'macroergic' compound, ATP, in photosynthesis two energy-rich compounds are formed, $NADPH_2$ and ATP, whose energy is then consumed for final sugar synthesis. The efficiency of the photosynthetic process alone, without losses due to imperfect light absorption by the green plant, is about 34 %, which is fairly good compared with solid-state or electrochemical photovoltaic devices. On the other hand, this is far from perfect energy transformation, a physicochemical property which laymen ascribe to biological systems. Nature does not need total accumulation of solar energy; the environment of the plants must also be heated by sunlight.

References

1. P. Mitchell, *Chemiosmotic Coupling and Energy Transduction*, Glynn Research, Bodmin, Cornwall, England, 1968.
2. P. Mitchell, *Science*, **206** (1979) 1148.
3. A. E. Senior, *Biochim. Biophys. Acta*, **301** (1973) 249.
4. See, for example, D. Osterhelt, *Mol. Subcell. Biol.*, **4** (1976) 133.
5. J. Barber, Ed., *Topics in Photosynthesis*, Vols. 1–3, Elsevier, Amsterdam, 1978.
   K. Sauer, *Accounts Chem. Res.*, **11** (1978) 257.
   P. Mitchell, *Biol Rev.*, **41** (1966) 445.
   H. T. Witt, 'Charge separation in photosynthesis', in: *Light-induced Charge Separation in Biology and Chemistry* (H. Gerischer and J. J. Katz, eds.), Verlag Chemie, Weinheim, 1979.
6. K. Sauer, *Accounts Chem. Res.*, **13** (1980) 249.

Appendix A

From M. Faraday, *Experimental Researches in Electricity,*
7th Series (1834).

The theory which I believe to be true expression of the facts of electro-chemical decomposition, and which I have therefore detailed in a former series of these Researches, is so much at variance with those previously advanced, that I find the greatest difficulty in stating results, as I think, correctly, whilst limited to the use of terms which are current with a certain accepted meaning. Of this kind is the term pole, with its prefixes of positive and negative, and the attached ideas of attraction and repulsion. The general phraseology is that the positive pole attracts oxygen, acids, etc., or more cautiously, that it determines their evolution upon its surface; and that the negative pole acts in an equal manner upon hydrogen, combustibles, metals and bases. According to my view, the determining force is not at the poles, but within the body under decomposition; and the oxygen and acids are rendered at the negative extremity of that body, whilst hydrogen, metals, etc., are evolved at the positive extremity.

To avoid, therefore, confusion and circumlocution, and for the sake of greater precision of expression than I can otherwise obtain, I have deliberately considered the subject with two friends, and with their assistance and concurrence in framing them, I purpose henceforward using certain other terms, which I will now define. The poles, as they are usually called, are only the doors or ways by which the electric current passes into and out of the decomposing body; and they of course, when in contact with that body, are the limits of its extent in the direction of the current. The term has been generally applied to the metal surfaces in contact with the decomposing substance; but whether philosophers generally would also apply it to the surface of air and water, against which I have effected electro-chemical decomposition, is subject to doubt. In place of the term pole, I propose using that of Electrode, and I mean thereby that substance, or rather surface, whether of air, water, metal, or any other body, which bounds the extent of the decomposing matter in the direction of the electric current.

The surface at which, according to common phraseology, the electric current enters and leaves a decomposing body, are most important places of action, and require to be distinguished apart from the poles, with which they are mostly, and the electrodes, with which they are always, in contact. Wishing for a natural standard of electric direction to which I might refer these, expressive of their difference and at the same time free from all theory, I have thought it might be found in the earth. If the magnetism of the earth be due to electric currents passing round it, the latter must be in a constant direction, which, according to present usage of speech, would be from east to west, or, which will strengthen this help to the memory, that in which the sun appears to move. If in any case of electro-decomposition we consider the decomposing body as placed so that the current passing through it shall be in the same direction, and parallel to that supposed to exist in the earth, then the surfaces at which the electricity is passing into and out of the substance would have an invariable reference, and exhibit constantly the same relations of powers. Upon this notion we purpose calling that towards the east the anode, and that towards the west the cathode; and whatever changes may take place in our views of the nature of electricity and electrical action, as they must affect the natural standard referred to, in the same direction, and to an equal amount with any decomposing substances to which these terms may at any time be applied, there seems no reason to expect that they will lead to confusion, or tend in any way to support false views. The anode is therefore that surface at which the electric current according to our present expression, enters: it is the negative extremity of the decomposing body; is where oxygen, chlorine, acids, etc., are evolved; and is against or opposite the positive electrode. The cathode is that surface at which the current leaves the decomposing body; and is its positive extremity; the combustible bodies, metals, alkalies, and bases, are evolved there, and it is in contact with the negative electrode.

I shall have occasion in these Researches, also, to class bodies together according to certain relations derived from their electrical actions; and wishing to express those relations without at the same time involving the expression of any hypothetical views, I intend using the following names and terms. Many bodies are decomposed directly by the electric current, their elements being set free: these I propose to call electrolytes. Water, therefore, is an electrolyte. The bodies which, like nitric or sulphuric acids, are decomposed in a secondary manner, are not included under this term. Then for electro-chemically decomposed, I shall often use the term electrolyzed, derived in the same way, and implying that the body spoken of is separated into its components under the influence of electricity: it is analogous in its sense and sound to analyze, which is derived in a similar manner. The term electrolytical will be understood at once: muriatic acid is electrolytical, boracic acid is not.

Finally, I require a term to express those bodies which can pass to the electrodes, or, as they are usually called, the poles. Substances are frequently spoken of as being electro-negative, or electro-positive, according as they go under the supposed influence of a direct attraction to the positive or negative pole. But these terms are much too significant for the use to which I should have to put them; for though the meanings are perhaps right, they are only hypothetical, and may be wrong; and then, through a very imperceptible, but still very dangerous, because continual, influence, they do great injury to science, by contracting and limiting the habitual views of those engaged in pursuing it. I propose to distinguish such bodies by calling those anions which go to the anode of the decomposing body; and those passing to the cathode, cations; and when I have occasion to speak of these together, I shall call them ions. Thus the chloride of lead is an electrolyte, and when electrolyzed evolves the two ions, chlorine and lead, the former being an anion, and the latter a cation.

Appendix B
Thermodynamics

Chemical reactions are mainly studied under constant pressure, usually standard atmospheric pressure (see Appendix D). The basic quantity for the energetics of chemical reactions[1] (as well as of other processes like solvation) is the reaction heat. The heat acquired or released by a reacting system between the end and the beginning of a chemical reaction is termed the enthalpy change, ΔH. The application of this concept is not restricted to chemical reactions, but this will not concern us here. ΔH depends only on the state of the system after the reaction and before it; the way in which the final state is reached is irrelevant (the Hess Law). Therefore, the enthalpy is included among the quantities of state (like a number of other quantities in thermodynamics). When the system accepts heat (the reaction is endothermic) the enthalpy is positive, while for heat release (an exothermic reaction) it is negative. When the reaction occurs to unit extent (cf. p. 72) the term reaction enthalpy, ΔH_r, is used. For many reactions, their reaction enthalpies have been tabulated.[2]

A further important thermodynamic quantity of state is the *Gibbs energy*. This quantity, connected with processes occurring at constant temperature and pressure, expresses the ability of the system to perform maximum useful work. This particular kind of work is almost exclusively connected with chemical reactions. When the volume is increased during a chemical reaction occurring at constant pressure, a definite (relatively small) amount of work is performed, for example the gases reacting in a cylindrical vessel closed by a piston will lift a weight placed on the piston. In a chemical reaction, however, a much larger amount of energy is usually required or released than that corresponding to the 'volume' work. A major part of this chemical energy can be transformed to 'non-volume' useful work, most frequently to electric work. The remaining part is the latent heat given by the product of the absolute temperature and another quantity of state, the entropy. This quantity is a measure of the disarray of the system (a crystal is more arrayed than a liquid which is more ordered than a gas, separated sorts of gases are more arrayed than their mixtures, etc.).

The Gibbs energy is defined by the equation

$$G = H - TS$$

where H is the enthalpy of the system, T the absolute temperature, and S the entropy. For the conditions after and before a reaction we can write

$$\Delta G = \Delta H - T\Delta S$$

where Δ denotes the difference in the value of a quantity of state after and before a reaction. For a reaction that occurs spontaneously, $\Delta G < 0$, while for a reaction occurring spontaneously in the opposite direction, $\Delta G > 0$. For an equilibrium, $\Delta G = 0$.

Quantity G is referred to the system as a whole. When the contribution of an individual component to this quantity is to be determined (regardless of the size of the system) G is differentiated with respect to the amount (in mols) of the given species n at constant temperature, pressure, and composition of the system. The resulting quantity is the chemical potential

$$\mu_i = \partial G / \partial n_i$$

References

1. K. Denbigh, *The Principles of Chemical Equilibrium, with Applications in Chemistry and Chemical Engineering*, Cambridge University Press, 1971.
   I. M. Klotz, *Chemical Thermodynamics*, Benjamin, New York, 1964.
2. See, for example, W. M. Latimer, *Oxidation States of Elements*, Prentice-Hall, New York, 1952.

Appendix C
Kinetics

In Appendix B the thermodynamics of reacting systems was discussed, i.e. their static properties. Now we shall be concerned with the dynamic aspect of reactions—the reaction kinetics. The basic quantity in this field, which can be expressed in thermodynamic terms, is the activation energy. When a chemical reaction occurs the molecules approach one another to a necessary distance (they 'collide'). Analogously, in an electrode reaction they approach the electrode/electrolyte solution boundary. However, not every collision is effective. The reacting partners (two particles, electrode and reacting particle, etc.) must possess sufficient energy (translational, vibrational, or rotational) to undergo the reaction. There is a definite boundary for the energy of reacting particles under which the reaction does not take place, termed the *activation energy*, $\Delta H^{\ddagger}$ (symbol H indicates that this is usually a kind of enthalpy). For the rate constant of a reaction the Arrhenius equation holds

$$k = k^0 \exp\left[-\Delta H^{\ddagger}/(RT)\right]$$

where k^0 is the pre-exponential term which depends only slightly on the temperature.

In order to derive an expression for the activation energy of an electrode reaction[1] (p. 98) a simplified model of the vibrating solvation sheath of the oxidized and of the reduced species must first be designed. Their condition will be characterized by the mean distance of solvating molecules from the centre of the ion, x. In fact, these molecules vibrate in a much more complicated mode, certainly not only in the direction to and from the ion; because there is a spectrum of vibrational frequencies, they also rotate, etc. For the sake of simplicity, a harmonic vibration will be assumed, yielding the potential energy of the solvation sheath in the initial state of the reaction

$$E_p(x) = \tfrac{1}{2}m\omega^2 (x - x_0^i)^2 + \alpha^i \tag{C.1}$$

and the potential energy in the final state of the reaction

$$E_p(x) = \tfrac{1}{2}m\omega^2 (x - x_0^f)^2 + \alpha^f \tag{C.2}$$

where ω is the angular frequency of vibration, m is the effective mass of the solvation sheath, x_0^i the equilibrium distance of solvating molecules from the centre of the ion in the initial, and x_0^f in the final, state of the reaction, and α^i and α^f arc the bonding energies of the solvating molecules in the initial and in the final

188

state of the reaction. All the quantities are referred to one ion, not to one mol of ions. The total energy of the ion-solvation sheath–electron system (either in the electrode or in the Fe^{2+} ion) consists of the potential energy E_p, the kinetic energy of the solvation sheath E_k, and the energy of the electron ε

$$E = E_p + E_k + \varepsilon \tag{C.3}$$

where $E_k = p^2/2m$, p denoting the impulse of all solvating molecules. The interaction between the ion and the electrode (for example, the adsorption) is supposed to be very weak so that the interaction energy will be neglected. Equation (C.3) generally describes the properties of the reacting system. The initial and the final state of the electron transfer reaction differ in the equilibrium distance of the solvation sheath x_0, in the bonding energy of the solvation sheath, and in the energy of the electron (initial value ε^i, final ε^f). A gross simplification is the assumption that neither impulse p nor vibration frequency ω change during the reaction. Furthermore, it is assumed that during the electron transfer no energy is exchanged between the system and its surroundings (such a process is termed *adiabatic*).

Under these conditions there is a definite distance of the solvation sheath from the centre of the ion, $x = x_r$, at which the energies of the initial and final states are equal

$$E_{(x_r)}^{(i)} = E_{(x_r)}^{(f)} \tag{C.4}$$

Substitution from (C.1) and (C.2) into (C.3) (separately for $E_{(x)}^{(i)}$ and $E_{(x)}^{(f)}$) and hence into (C.4) yields the equation for the critical distance, x_r

$$x_r = \frac{\alpha^f + \varepsilon^f - \varepsilon^i - \alpha^i + \frac{1}{2} m \omega^2 (x_0^{f2} - x_0^{i2})}{m \omega^2 (x_0^f - x_0^i)} \tag{C.5}$$

The dependence of the energy of the reacting system for the reactant and for the reaction product on the distance of the solvation sheath from the ion centre x is depicted in Fig. 110. Distance x_r corresponds to the intersection of the curves for $E_{(x)}^{(i)}$ and $E_{(x)}^{(f)}$. During the reaction the energy of the system first follows the curve for $E^{(i)}$ and then that for $E^{(f)}$. At the point $x = x_r$ the energy of the system is given by the equation

$$E_r = \frac{p^2}{2m} + \frac{1}{2} m \omega^2 (x_r - x_0^i)^2 + \varepsilon^i + \alpha^i \tag{C.6}$$

The activation energy of the elementary process (with the participation of individual particles) $E^{\ddagger}$ is given by the difference of the energy at point $x = x_r$ and the energy of the initial state; consequently

$$
\begin{aligned}
E^{\ddagger} = E_{(x_r)}^{(i)} - E_{(x_0)}^{(i)} &= \frac{1}{2} m \omega^2 (x_r - x_0^i)^2 \\
&= \frac{(E_r + \varepsilon^f - \varepsilon^i + \alpha^f - \alpha^i)^2}{4E_r}
\end{aligned} \tag{C.7}
$$

The difference $\varepsilon^f - \varepsilon^i$ refers to the state of the electron after and before the reaction. Quantity ε^f can be identified with the ionization potential of the reduced

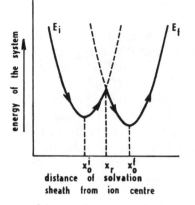

Fig. 110. Change in the energy of the reacting system during the electron transfer process

form I/N_A multiplied by -1 (cf. p. 15). The initial state of the particle is characterized by the energy of the Fermi level, and consequently by the electrochemical potential of the electron in the metal. $\varepsilon_F = \tilde{\mu}_e/N_A$. This quantity is a sort of Gibbs energy while ε^f is a sort of enthalpy. In order to transform the Fermi level energy to an enthalpy term, quantity $q(q = Ts_e$, where s_e is the entropy of the electron) must be added with the result

$$\varepsilon^i = \varepsilon_F + q = \mu_e^0/N_A - e\varphi + q \tag{C.8}$$

where $e = F/N_A$ is the electron charge. The difference between the bonding energies $\alpha^f - \alpha^i$ can be identified with the difference in the solvation energies of the final and initial states of the ion

$$\alpha^f - \alpha^i = (\Delta H_s^f - \Delta H_s^i)/N_A \tag{C.9}$$

In this way the activation energy of the cathodic process is obtained

$$\Delta H_{red}^{\ddagger} = E^{\ddagger} N_A = \frac{[E_r N_A - (I + \mu_e^0 + T\Delta S_e + \Delta H_s^f - \Delta H_s^i - F\varphi)]^2}{4E_r N_A} \tag{C.10}$$

When a considerable larger value of $E_r N_A$ than of the expression in brackets is assumed (C.10) can be simplified to (cf. (2.54))

$$\Delta H_{red}^{\ddagger} \approx E_r N_A/4 - \tfrac{1}{2}(I + \mu_e^0 + T\Delta S_e + \Delta H_s^f - \Delta H_s^i + F\varphi) \tag{C.11}$$

It holds for the rate constant of the cathodic process ($\varphi = E + $ const) that

$$k_{red} = k'' \exp(-\Delta H^{\ddagger red}/RT) = k' \exp(-EF/2RT)$$

which is identical to the equation on p. 94 for $\alpha = \tfrac{1}{2}$ and $k' = k^0 \exp[\alpha FE^0/(RT)]$.

The simplification leading to equation (C.11) is not always justified. In the Fe^{3+}/Fe^{2+} system the dependence of k_{red} on the electrode potential has a curvilinear shape (Fig. 111) showing that the unsimplified equation (C.10) has to

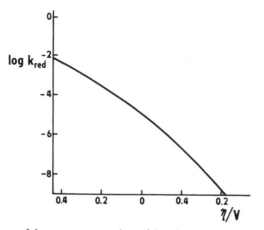

Fig. 111. Dependence of the rate constant k_{red} of the electrode reaction $Fe^{3+} + e = Fe^{2+}$ at a platinum electrode on the overpotential. Electrolyte: 2.25 m M Fe^{3+}, 2.25 m M Fe^{2+}, 0.5 M $HClO_4$. According to J. Weber, Z. Samec, and V. Mareček

be used for analysis of the system. Simple relations of the type (2.38) and (2.39) with constant α can then be applied to processes taking place only in the neighbourhood of the standard potential as an approximation.

References

1. R. A. Marcus, *Ann. Rev. Phys. Chem.*, **15** (1964) 155.
   V. G. Levich, 'Present state of the theory of oxidation–reduction in solution bulk and electrode reactions', in: *Advances in Electrochemistry*, Vol. 4 (P. Delahay and C. W. Tobias, eds.), Wiley–Interscience, New York, 1966, p. 249.
   W. Vielstich and W. Schmickler, *Elektrochemie II—Kinetik elektrochemischer Systeme*, Steinkopff, Darmstadt, 1976, p. 32.

Appendix D
SI Units

The name International System of Units (SI) has been adopted by the Conference Generale des Poids et Mesures for the coherent system based on the units: metre, kilogramme, second, ampere, kelvin, and candela. In the International System there is one and only one basic unit for each physical quantity. This is either the appropriate basic SI unit itself or the appropriate derived SI unit formed by multiplication and/or division of two or more basic SI units. A few such derived SI units have been given special names and symbols.

Decimal fractions and decimal multiples both of the basic and of the derived SI units may, however, be constructed by use of approved prefixes.

The definition of the basic SI units of importance in electrochemistry is:

metre: the metre is the length equal to 1650763.73 (exactly) wave-lengths in a vacuum of the radiation corresponding to the transition between the energy levels $2p_{10}$ and $5d_5$ of the pure nuclide ^{86}Kr.

kilogramme: the kilogramme is the mass of the International Prototype Kilogramme which is in the custody of the Bureau International des Poids et Mesures at Sevres, France.

second: the second is the duration of 9192631770 periods of the radiation corresponding to the transition between the two hyperfine levels ($F = 4, M_F = 0$, and $F = 3, M_F = 0$) of the fundamental state ($2S_{\frac{1}{2}}$) of the atom of caesium 133.

ampere: the ampere is that constant current which, if maintained in two parallel rectilinear conductors, of infinite length and of negligible circular cross-section, at a distance apart of 1 metre in a vacuum, would produce a force between the conductors equal to 2×10^{-7} newton per metre of length.

kelvin: the kelvin, unit of thermodynamic temperature, is the fraction 1/273.16 exactly of the thermodynamic temperature at the triple point of water.

mol: the mol is an amount of substance in a system which contains as many elementary units as there are carbon atoms in 0.012 kg (exactly) of the pure nuclide ^{12}C. The elementary unit must be specified and may be an atom, a molecule, an ion, an electron, a photon, etc., or a specified group of such entities.

It should be noted that the current expression 'number of mols' does not often refer to a plain number but just the 'amount of the substance in mols', having thus the dimension mol. For example, the ratio V/n, where V is the volume of a homogeneous system (in cm^3) and n the number of the substance in mols (the so-called 'number of mols'), is called the molar volume. It has the dimension $cm^3.mol^{-1}$.

With the exception of ampere, the other electrical units used in electrochemistry are derived ones. For example, the main unit of charge is the coulomb (C) which is equal to the charge passed through the cross-section of a conductor while a current of 1 A has been flowing for 1 s. The other units of importance to electrochemistry are:

potential	volt	$V \ (= W.A^{-1})$
resistance	ohm	$\Omega \ (= V.A^{-1})$
conductance	siemens	$S (=\Omega^{-1})$
capacitance	farad	$F \ (= C.V^{-1})$

The derived unit for pressure is the pascal (Pa) which is connected to the former unit, atmosphere, by the relation 1 at $= 101325$ Pa. This 'standard atmospheric pressure' is used in definitions of standard states of substances and systems. We have become acquainted with this way of standardization in the case of the standard hydrogen electrode. In order to avoid the necessity of putting the value of $p_0 = 101325$ Pa into various expressions of the type of equation (2.15), it is advantageous to use the dimensionless relative pressure, $p' = p/p_0$, whenever it is possible.

The permittivity of a dielectric is given by the equation

$$\varepsilon = D\varepsilon_0$$

where D is the relative permittivity (dielectric constant) and ε_0 the permittivity of a vacuum ($\varepsilon_0 = 8.85 \times 10^{-12}$ F.m^{-1}). On defining the permittivity in this way we have to write the equation for Coulomb's law

$$F = \frac{Q_1 Q_2}{4\pi D\varepsilon_0 r^2}$$

where F is the force (in newtons) and Q_1 and Q_2 the point charges (in coulombs) present in a vacuum a distance r apart (in metres).

Index